Experimental Investigation and Optimization of Aluminum Matrix Composites

Niver Keiv

TABLE OF CONTENT

CHAPTER 1

INTRODUCTION

1.1 BACKGROUND

Materials have an important role in the development of human living standards. They are the basics of all engineering applications. The importance of metals is established by the fact that the human eras are named after the commonly used materials of that time such as the Stone Age, Bronze Age and Iron Age. With the commencement of the Stone Age, humans are consistently reforming materials and tailoring their properties according to the demand of modern-day technology. The development of technology depends on the material. The advancement of every field of engineering relies upon materials. Recent decades have seen a growing demand for materials with higher strength-to-weight ratios, greater stiffness, higher working temperatures, excellent wear resistance and tailorability. Aircraft, automotive and many other industries have an ever-increasing demand for more advanced materials for innovative technology [1, 2].

1.2 COMPOSITE

In a composite, two or more ingredients are synthesized macroscopically which are not dissolved in each other. An embedding component is known as the matrix phase and a reinforcing component is known as the reinforcing phase. There are a variety of ways to use reinforcing phase material, such as fibers, flakes and particles. The matrix phase of materials usually consists of a continuous phase. The two phases are embedded in each other in such a way that they have a recognizable interface between them [3].

A couple of last decades have shown that composites are better engineering materials than conventional metals. They have an excellent strength-to-weight ratio, better stiffness, large working temperature, superior resistance to wear and they also provide the convenience of modifying these properties for a specific purpose. The tailoring of properties of the composite material makes it popular in all industries. In many applications, only composite material provides a unique synthesis of properties that cannot be fulfilled by other materials [4, 5].

1.2.1 Components of a Composite Material

A composite consists of two components, one is the matrix phase, the other is the reinforcement phase, which is mixed to tailor the matrix properties. The components of the composite are as follows:

1.2.1.1 Matrix

In composite materials, the matrix forms a regular phase of composite material, binding and encapsulating the reinforcement to create a unified structure. They provide structural integrity, load transfer capabilities and environmental protection for the composite. The properties of the matrix material, such as its mechanical strength, flexibility, thermal stability and chemical resistance, significantly influence various properties of the composite.

The selection of the appropriate matrix material relies upon factors such as the needed characteristics of the composite, the application requirements and the compatibility with the reinforcement material. The matrix material should be suitable for the reinforcement in terms of adhesion, mechanical properties and thermal expansion characteristics to ensure effective load transfer and overall composite performance [5].

1.2.1.2 Reinforcement

In composite materials, reinforcement refers to the addition of strong, stiff materials to alter the characteristics of the matrix material as per the requirement. There is a wide variety of reinforcements used in composites and the characteristics of the composites is highly affected by the reinforcements. The various characteristics of reinforcement are due to the variety of orientations and multiple ways of their assembly. They can be in the shape of particles, whiskers, fibers or sheets [5].

1.2.2 Classification of Composites

Composites can be categorized according to different factors, including the properties of both the matrix and reinforcement materials. Generally, composites can be categorized on the basis of the type of reinforcement and the matrix material. Following are the classifications based on these bases:

1.2.2.1 Classification based on reinforcement type

Categorized by their reinforcement type, composites fall into two main groups: those reinforced with fibers and those reinforced with particles. In the case of fiber-reinforced composites, two additional distinctions emerge: discontinuous fiber and continuous fiber composites. Fig. 1.1 represents different types of composites based on the reinforcement type.

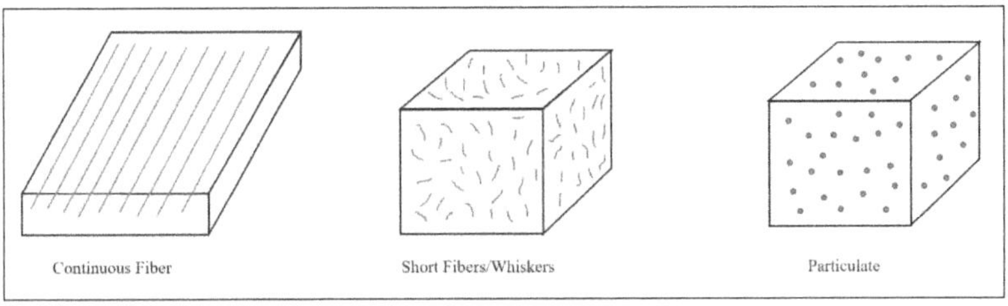

Fig. 1.1 – Different types of reinforcements [4]

Particle Reinforced Composites utilize particles (including platelets) as reinforcements. These particles can have any shape, such as spherical, cubic, tetragonal, or platelet like and may exhibit regular or irregular forms. To enhance matrix material performance, particle fillers are incorporated. These fillers serve to modify attributes like hardness, wear, abrasion resistance, strength and thermal conductivity. Achieving desired enhancements in the properties of a composite requires meticulous consideration of factors like particle type, size, shape and concentration during the selection and design process. Optimization of these parameters is essential to modify the composite material to meet specific performance requirements, ensuring that the desired improvements in its properties are effectively realized [6].

Fiber reinforced composites are materials made by embedding fibers of high strength and stiffness into a matrix, typically a polymer, to improve the properties. Fibers are characterized by their significantly greater length compared to the cross-section. The ratio of reinforcement has a substantial impact on its capacity to convey its properties to the material. Fibers perform a vital role in improving the fracture resistance of the matrix. Their elongated shape inhibits the propagation of cracks perpendicular to the reinforcement, thereby mitigating the risk of failure, particularly in matrixes prone to brittleness [7].

Fibers produced from non-polymeric materials have exceptionally excellent strength throughout length due to the minimal presence of large flaws. Polymeric materials fibers are strong and stiff because of their molecular orientation, but due to their small cross-sectional area, they cannot be used effectively in engineering operations. The matrix functions as a cohesive agent, binding the fibers, transferring loads to them and shielding them from potential damage. In contrast to continuous fiber composites, the matrix assumes a more pronounced role in facilitating load transfer in discontinuous fiber composites [8].

1.2.2.2 Classification based on type of the matrix

There are three major classes of composites under this classification. Here are the following:

Polymer matrix composites (PMCs) are advanced materials comprising a polymer base (such as epoxy, polyester, or urethane) reinforced with fine fibers of materials like graphite, aramid, boron etc. PMCs are typically chosen based on their characteristics, like mechanical strength, stiffness, thermal stability and environmental resistance, to meet the peculiar requirements of the material for a given application. Common types of polymer matrices used in composites include epoxy, polyester, vinyl ester, polyimide and thermoplastic polymers such as polypropylene and polyethylene. Each type of polymer matrix has its unique set of properties that can be tailored to suit the desired performance characteristics of that composite material [9].

The polymer matrix in a PMC acts as a binder, holding the reinforcing materials together and transferring applied loads between them. It additionally shields the reinforcing materials from external elements like humidity, chemicals and ultraviolet (UV) radiation. The polymer matrix can be tailored to provide the desired balance of properties which includes strength, stiffness, toughness and thermal stability [10].

Ceramic Matrix Composites (CMCs) contain ceramics as the matrix material, which is reinforced with short whiskers or fibers like carbon fibers, silicon carbide etc. Alumina (Al_2O_3) and silicon carbide (SiC) ceramic matrix composites (CMCs) are commonly employed in situations characterized by elevated temperatures and environments where chemical susceptibility is the primary consideration. Due to their inherent weaknesses in tension and shear, ceramics are often used as reinforcement in the form of particles, such as zinc and calcium phosphate [11].

Metal matrix composites (MMCs) are composite materials containing metallic or ceramic reinforcement particles dispersed in metal or alloy matrix materials. MMCs are considered superior to ceramics and polymer matrixes due to excellent toughness, withholding properties at high temperatures, better strength, enhanced electrical and thermal conductivity and more erosion resistance. MMCs have gained global recognition as compared to wrought alloys because of their remarkable strength, wear resistance, stiffness as well as creep resistance. Metallic composites, including aluminium, titanium, magnesium and nickel-chromium alloys, exhibit high toughness and impact strength due to their ability to undergo plastic deformation under impact. [12].

MMCs are reinforced with different kinds of reinforcement such as particles, fibers, or whiskers. MMCs reinforced with particles of a different material are called particle reinforced metal matrix composites. These particles, such as ceramics, carbides, or oxides, are scattered inside the metal matrix to improve its various properties. It offers a unique combination of characteristics, including improved strength, wear resistance, stiffness and thermal stability. The reinforcement particles act as barriers to dislocation movement, providing increased strength and hardness. They also help to reduce thermal expansion and improve the overall thermal conductivity of the material [13]. Table 1.1 represents the various possible uses of MMCs in manufacturing different parts of automobiles.

Table 1.1 - Possible uses of MMCs in the automotive industry [6]

System	Component	Properties required
Engine	Piston Crown, Valve	Good resistance to fatigue, creep and wear at High temperature
	Piston Ring Groove	Good wear resistance, weight reduction
	Rocker Arm	Weight reduction and high stiffness and wear resistance
	Wrist Pin	Good resistance to wear and creep with high specific stiffness
	Cylinder Block (liner)	Good wear and seizure resistance, low friction, low weight
	Connecting Rod	Good specific stiffness and low weight
	Bearings	Low weight and reduced friction
Suspension	Struts	Good damping capacity with high stiffness
	Shift Forks	Good wear resistance with low-weight
	Drive Shaft	Good specific stiffness and fatigue resistance
	Gears, Wheels	Good wear resistance with low-weight
Housings	Bearing and Pumps	Good wear resistance with low-weight
Brakes	Disk Rotors, Break Drum, Callipers	Good wear resistance with low-weight

1.3 ALUMINIUM MATRIX COMPOSITE

An aluminium matrix composite (AMC) refers to a type of MMC that consists of aluminium as the matrix or base material and one or more additional components added to enhance its properties. The additional components, known as reinforcements or fillers, are typically stronger and stiffer materials such as ceramic particles, fibers, or nanoparticles. AMCs are designed to combine the lightweight nature of aluminium with the better form of mechanical, thermal, tribological and physical properties provided by reinforcements. In AMCs, the matrix material provides exceptional resistance to corrosion, high levels of electrical and thermal conductivity and favourable formability. By incorporating reinforcements, the resulting composite can exhibit enhanced strength, stiffness, wear resistance and elevated temperature capabilities [14].

A combination of reinforcement materials in AMCs will be determined by its specific application requirements. For example, ceramic particles like silicon carbide (SiC) or alumina (Al_2O_3) can be added to improve hardness and wear resistance. Fibrous reinforcements such as carbon fibers or silicon carbide fibers can enhance strength and stiffness. Nanoparticles like graphene or carbon nanotubes can be used to further improve mechanical properties at the nanoscale level. AMCs find applications across diverse industries, including automotive, aerospace, electronics and sporting goods. In the automotive industry, AMCs can be utilized for engine parts, suspension components and body panels to reduce weight while maintaining strength and safety. In aerospace, AMCs can be used for lightweight structural components in aircraft, leading to improved fuel efficiency. Additionally, AMCs have been explored for electronic packaging, heat sinks and other high-temperature applications [15].

1.4 PROCESSING TECHNIQUES OF AMCS

The processing techniques for AMCs are crucial for achieving the desired properties and performance. Processing techniques may be categorized into two primary sections: solid-state processing and liquid state processing.

1.4.1 Solid State Processing Routes

Diffusion bonding stands out as a frequently utilized method for joining metals, whether they are similar or dissimilar in nature. This technique relies on the inter-diffusion of atoms from two metallic surfaces that come into contact at elevated temperatures. It offers flexibility as it can be utilized for a broad spectrum of metal matrix, leading to precise control over factors such as fiber orientation and volume fraction. However, diffusion bonding does have some downsides. The process tends to be time-consuming due to the requirement for extended processing durations. Additionally, it requires high temperature and pressure, which can increase production costs. Furthermore, diffusion bonding may not be the ideal choice for the fabrication of components with complex shapes due to its inherent limitations [4, 16].

Powder blending and consolidation offer a versatile approach to manufacturing AMCs. A metal alloy particle can be amalgamated with particles or short fibers of ceramic. Following the blending stage, a sequence of steps is undertaken, including compression in a cold state, canning, degassing and ultimately a high-temperature consolidation process. During this process, matrix material undergoes significant deformation due to the presence of rigid particles or fibers. As a result, the AMCs fabricated through this process contain a thin oxide layer that assumes a plate-like structure. These oxide particles serve as dispersion strengthening agents, exerting an influence on the properties of the base alloy, unusually in the process of heat treatment [16].

The Physical Vapor Deposition (PVD) technique entails the deposition of a metal in the shape of a thin layer coating. This is the result of the condensation of vaporized metals onto a designated surface. During the production of AMC, a high-intensity electron beam is utilized as the heat source. This electron beam generates the vapor of the desired material, which is then allowed to condense onto the surface of the fibers [16].

1.4.2 Liquid State Processing Routes

The stir casting method also known as the vortex method, is a broadly used approach for fabricating composites. The workflow diagram of the stir casting technique is shown in Fig. 1.2. This method involves adding reinforcing phases in powder form into molten metal while mechanically stirring the melt. The stir casting method involves vigorous stirring of the molten matrix material to formulate a vortex at the upper layer of the molten matrix material, through which the reinforcement material is inserted. Stirring continues until its melt is cast into the desired mould. This technique facilitates the transfer of particles into the molten metal and keeps them suspended during composite fabrication [17, 18].

Stir casting is a preferred method for fabricating MMCs due to its simplicity, cost-effectiveness, flexibility in choosing different types and sizes of reinforcements and suitability for large-scale production. Stir casting enables uniform distribution of reinforcement within the matrix, leading to enhanced mechanical properties, unlike powder metallurgy or chemical vapor deposition methods. Additionally, the process is relatively quick and requires less specialized equipment, which reduces manufacturing costs and complexity [19].

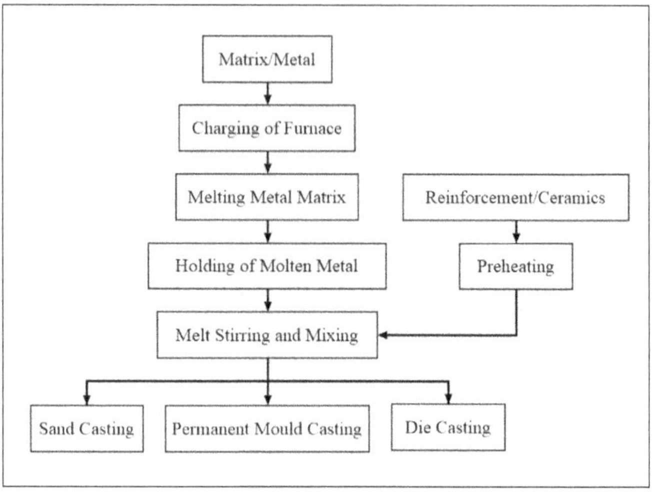

Fig. 1.2 Stir casting method

Spray deposition is a method that involves the atomization of molten material, typically metals, leading to the formation of solidified droplets in powder form. These semi-solid droplets are collected onto a substrate. The process is rapid, with a quick transition from liquid to solid phase, followed by slow cooling in the solid state. This gradual cooling results in finer grains, although it doesn't significantly increase solute solubility. AMCs produced through spray deposition exhibit an uneven distribution of ceramic particles but possess strong interfacial bonding without any interfacial reactions. Typically, there is about 5% porosity in the as-sprayed state, but secondary processing can eliminate this porosity. The cost of producing AMCs using this technique falls between that of stir casting and powder metallurgy processes [20, 21].

Squeeze casting, also known as Infiltration processing, stands as one of the traditional methods utilized in casting materials essential for the automotive industry. This process involves the preparation of porous preforms with fiber or particulate reinforcements, followed by the infiltration/injection of molten metal under a high pressure of approximately 100 MPa. The aim is to fill all the pores and channels of the preforms. The advantage of this technique lies in its short processing time, minimizing the interaction between the reinforcement and molten metal. Consequently, composite defects such as porosity and shrinkage cavities are notably absent [22]. The rapid liquid-solid phase change in this method leads to non-equilibrium solidification, resulting in a fine grain structure for the composite. In addition to achieving good distribution and strong interfacial bonding, the technique ensures a smooth surface finish for the product. However, it comes at a significant cost due to the need for expensive tools capable of withstanding high pressure and temperature. Moreover, the maintenance of these tools adds to the overall product cost. Damaged tools can introduce errors in product dimensions [23, 24]. Fig. 1.3 present a comprehensive comparison of costs, which outlines the expenses associated with liquid state, spray forming, powder metallurgy and diffusion bonding techniques.

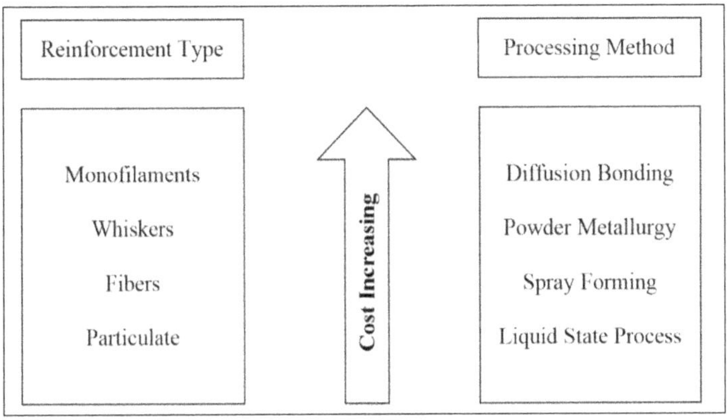

Fig. 1.3 Cost comparison of different manufacturing processes

1.5 WEAR

Wear can be explained as the deterioration or depletion of substances from surfaces of solids that come into contact during actions such as sliding, rolling or impact. Typically, wear arises from the interplay between the microscopic roughness of the surfaces. As they move relative to each other, these rough points, known as asperities, undergo deformation or detachment. This can result in the creation of wear debris or their transfer to the opposite surface. It's crucial to emphasize that wear is not an intrinsic characteristic of the material, instead, it relies on the particular circumstances of the surfaces involved and the operational variables in play [25].

The definition of wear is commonly based on material loss from one or all the contacting surfaces. American Society of Lubrication Engineers (ASLE) describes wear as "removal of material by mechanical action" [26]. Hence, wear stands as a surface-related occurrence brought about through material displacement and detachment. Contrary to situations involving complete structural failure, wear concerns manifest as a gradual reduction in material mass, causing dimensional modifications in components over extended periods. [27].

1.5.1 Types of Wear

Wear has several types, each categorized based on the specific mechanisms and conditions causing the deterioration. The main types of wear include:

Adhesive wear arises when two nearly flat solid objects come into contact and slide against each other, either without any lubrication or with some lubrication present. As these objects slide, small high points on their surfaces, known as asperities, come into contact and are gradually worn away. This process can result in tiny bits of material detaching from one surface and sticking to the other surface. These detached fragments might move between the surfaces multiple times during contact or could be dislodged as loose particles. Many researchers have explored the different ways in which these material fragments are transferred from one to another surface [28].

Abrasive wear happens when a harder covering move over a softer one. If the material is ductile, the contact area gets deformed, while brittle materials might fracture. Abrasive wear can be either two or three-body. In two-body, one of the rubbing surfaces is hard. In three-body, a hard third body, like small abrasive particles, gets caught between surfaces during sliding. These particles, being hard, can lead to wear on one or both surfaces. Typically, the initial wear is adhesive, producing particles that can lead to three-body abrasive wear if they become lodged between surfaces. [29].

Surface fatigue wear emerges from dynamic loading while sliding or rolling. Periodic loading and unloading cycles cause cracks in the surface or subsurface. Over cycles, materials may fail due to surface damage, forming larger pits. Fatigue results in minimal wear before failure, so weight loss is not a significant factor. Instead, the number of cycles until fatigue sets in is the most critical consideration. Rolling fatigue failure time depends on shear stress amplitude, lubrication and material properties. As sliding occurs, deformation leads to the formation of cracks on the surface or subsurface, which propagate with the load and result in surface fatigue failure [30, 31].

Impact wear can be categorized into erosive and percussive wear. Erosive wear happens when solid particles and liquid droplets strike a surface, along with the rupture of fluid-formed bubbles [32]. Percussive wear arises from repetitive impacts of solid objects, gradually causing material loss. Similar to adhesive and abrasive wear, erosive wear can induce plastic deformation or fracture. Brittle materials erode as cracks form and intersect, radiating from the point of impact by eroding particles whereas cutting and deformation mechanisms are responsible for material removal in ductile materials [33, 34]. Cavitation erosion happens when bubbles formed during the movement of fluid against a solid surface collapse upon impact [35].

Chemical or corrosive wear arises when there is a motion like sliding, rolling or rotation in a corrosive environment. Corrosion happens due to chemical or electrochemical reactions between the interface and surroundings. Oxygen, especially in the air, is a potent corrosive medium, leading to oxidative wear. When sliding is limited, corrosion may slow due to a surface layer formed by corrosion products. However, sliding can remove this layer, accelerating corrosion. So, both a chemical setting (corrosion) and friction are necessary for chemical wear. This wear type is common in auto parts like brakes and engines [36, 37].

Electrical-arc-induced wear occurs due to a strong electric potential develops across a thin air film during sliding. This causes dielectric breakdown and creates arcing. Intense arcing in a short span melts and then re-solidifies the heat-affected zone. This process could lead to phase changes and hardness alterations. High-density arcs might generate significant craters that can be fractured or sheared during sliding or oscillating movement. These resulting fragments can contribute to three-body abrasion. The occurrence of arcing can even trigger a combination of wear modes, possibly culminating in the catastrophic failure of the component [38, 39].

Fretting wear can occur when slight back-and-forth motion happens between two surfaces in contact. This motion breaks the protective film on the surface, leading to the oxidation of exposed metal and the creation of damaging wear debris. Fretting wear is a type of wear where normal load causes asperities to adhere, while oscillation leads to fractures, producing wear debris. When combined with corrosion, fretting is termed fretting corrosion [40, 41].

1.5.2 Stages of Wear

The process of wear is typically measured using wear rate, which represents the mass or volume detached per unit time. The wear rate is not constant and varies over time. It can show different patterns, starting low and increasing later, or the opposite, depending on the situation. Wear rate depends on material properties, surface conditions and operational factors. The wear-time curve is shown in Fig. 1.4. The wear curve can be classified into regions as follows (i) running-in, (ii) steady-state and (iii) breakdown. In the running-in phase, material properties and surface conditions significantly influence wear.

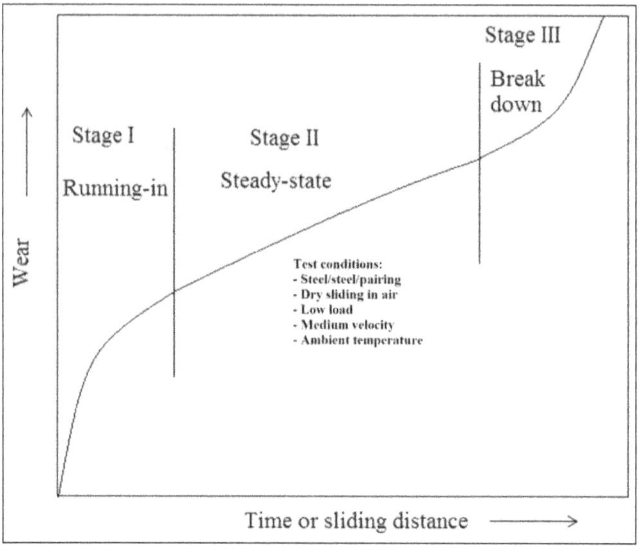

Fig. 1.4 - Wear - time curve [43]

Here, wear loss increases progressively with time or sliding distance. During the steady-state phase, the wear rate remains relatively constant, while in the breakdown phase, mass loss may increase exponentially until failure. Linear wear loss over time is common in abrasive wear, while sliding wear is influenced by adhesion, surface fatigue, or tribo-chemical reactions, leading to the observed three-phase wear patterns [42, 43].

1.5.3 Theories of Wear

Various wear mechanisms, including adhesion, abrasion, erosion and corrosion can occur during wear. The various theories applicable to these wear mechanisms are as follows:

1.5.3.1 Classical wear theory

Rabinowicz while studying wear patterns observed that local plastic deformation of contacting asperities at sufficiently high pressure causes adhesion at contact junctions of mating surfaces. The hardness of the materials in contact performs a primary job in deciding the actual area of asperity connection. Hence, hardness at contact points is of prime importance [44]. However, the effect of the material microstructure was completely ignored, hence, the theory remains limited only to ideal conditions without taking into consideration the nucleation and growth of the crack. Material removal was solely attributed to the cohesive failure of asperities. Archard developed the subsequent equation to quantify the volume loss attributed to wear,

$$\Delta V = \frac{k \times SD \times W}{H} \qquad \textbf{(1.1)}$$

In Eq. (1.1), k represents the wear coefficient, SD denotes the sliding distance, W signifies the applied weight and H stands for the bulk hardness [45].

1.5.3.2 Oxidative Wear Theory

Oxidation wear theory is a concept that explains the process of wear in materials due to the chemical reaction between a solid surface and oxygen in the presence of elevated temperatures. This theory is particularly relevant in high-temperature environments where materials are exposed to oxidative conditions. Oxidation wear occurs when the surface of a material reacts with oxygen to form oxides, which can lead to the degradation and loss of material [46]. oxidation wear theory is crucial for designing materials and components that can withstand high-temperature and oxidative environments, such as those found in aerospace, power generation and manufacturing industries. It helps engineers and researchers develop strategies to minimize wear and extend the service life of critical equipment and machinery [47].

1.5.3.3 Delamination Wear Theory

Suh has given delamination wear theory for low sliding velocities with the help of wear debris formed [48]. Adhesive, fretting and fatigue wear can be explained with this theory. This theory can be explained under following points-

- Adhesion and ploughing are responsible phenomena for load transmission during sliding which may either deform or remove asperities in contact due to repeated loading. This creates a relatively smooth surface and a wider area is available for contact for harder asperities during cyclic loading.
- Iterative loading leads to the build-up of shear plastic deformation, predominantly induced by rigid irregularities on a more yielding surface.
- Crack initiation occurs beneath the surface, driven by subsurface deformation induced by highly compressive tri-axial stresses located just beneath the contact regions.
- As loading persists, cracks continue to propagate along with deformation, eventually merging with neighbouring cracks. The direction of crack propagation tends to align along the surface, with the specific deepness influenced by the characteristics of the material and the coefficient of friction.

1.5.3.4 Quinn's Theory

According to his theory, excessive frictional heating at high velocities causes temperature rise and a thick oxide layer is formed on a sliding surface. This oxide layer is removed after a critical thickness leading to wear. The oxide film formed in the beginning is worn away leading to severe wear, but over a period of time worn surface recovers by an unknown process and the wear rate decreases drastically. But after a critical thickness, the oxide film becomes weak enough to withstand load and frictional shear stress weakens the film and it is detached as debris [49, 50].

1.5.3.5 Tribolayer theory

Jiang and other authors conducted a comprehensive review of dry sliding wear mechanisms. They noted a shift from severe to mild wear after a specific sliding duration, attributing this transition to the development of a stable tribolayer. The tribolayer is created through the transfer of debris from one surface to another [51]. Particles of plastically distorted and oxidized debris amalgamate to form a resilient protective layer, which effectively reduces the wear rate. This phenomenon becomes more pronounced at elevated temperatures, resulting in the development of a smooth tribolayer or a mechanically mixed layer (MML) that may incorporate particles from both surfaces. Material transfer occurs repeatedly during this process to generate debris. The critical thickness of the transfer layer varies with sliding conditions, but the debris particles are primarily produced through delamination. It is important to note that the formation of the tribolayer is a phenomenon not considered in both classical adhesive theory and delamination theory of wear [52, 53].

1.6 FRICTION

Friction is defined as the resistance encountered when a solid body moves tangentially in relation to another contacting body, acting in opposition to the motion. This resistance can manifest as either dry friction or friction by any fluid. Friction in the dry state is often referred to as "Coulomb" friction and arises between two dry surfaces that are inclined to move relative to each other. In contrast, fluid friction occurs between adjacent layers within a fluid, where these layers move at varying velocities relative to each other. This phenomenon is typically observed in the presence of liquid or gaseous lubricants positioned between bearing surfaces. The initial understanding of friction dates back to the fifteenth and sixteenth centuries, with contributions from Leonardo da Vinci and Guillaume Amontons. They introduced two fundamental principles of friction:

- No correlation exists between friction and the apparent contact area.
- Friction exhibits a direct increase to the normal load.

However, later on, Coulomb made significant advancements in the study of friction. He clearly distinguished between static and kinetic friction and introduced a third law, stating that friction is independent of sliding velocity. It became evident that friction is more influenced by the response of the contacting materials within a system rather than just their inherent properties. Moreover, operating conditions and interface characteristics are key considerations as well. It is worth noting that metals with clean, uncontaminated surfaces tend to exhibit a high coefficient of friction, whereas surfaces tainted with contaminants exhibit weaker adhesion and a lower coefficient of friction [54, 55].

1.6.1 Theory of Friction

Multiple factors contribute to the total friction experienced between surfaces in contact. These factors include adhesion between the mating surfaces, the interaction of surface asperities, ploughing of one surface by another, surface deformation and the entrapment of wear particles [56, 57]. The adhesion and deformation play pivotal roles in friction. They proposed that interacting asperities undergo plastic deformation, reaching a pressure equal to the hardness of material (H). Consequently, the actual contact area, denoted as 'Ar,' can be calculated by Eq. (1.2):

$$A_r = \frac{W}{H} \qquad \textbf{(1.2)}$$

Where W is the applied load. Assuming that metals weld together over a field of association and the tangential force (F) required for relative motion is equivalent to the multiplication of shear strength (S) and real field of association 'Ar'. Thus, the expression for the COF (μ) is given by Eq. (1.3),

$$\mu = \frac{F_f}{F_n} = \frac{A_r S}{A_r H} = \frac{S}{H} = \frac{\text{Shear strength of junctions}}{\text{Indentation hardness of softer material}} \qquad \textbf{(1.3)}$$

In the above equation, F_f represents the frictional force and F_n represents the applied normal force in Newton, H denotes the flow hardness measured in N/m^2 and S indicates the shear strength expressed in N/m^2.

Two primary elements influencing friction during sliding are adhesion and ploughing. The third body wear particle entrapped between two surfaces contributes in a major way to overall friction. The following factors contribute to the overall coefficient of friction:

- Bonding of the flat areas amid the sliding surfaces (μ_a)
- Cutting through hard asperities and abrasion caused by wear particles (μ_{plow})
- Surface asperities undergoing deformation (μ_d)
- Entrapment of wear particles (μ_{part})

$$\mu = \mu_a + \mu_{plow} + \mu_d + \mu_{part} \quad \textbf{(1.4)}$$

The final value of coefficient of friction is given by Eq. (1.4). The relative contribution of the above-mentioned factors depends on the sliding conditions, type of materials and the environment during the wear process [33].

1.6.2 Stages of Friction

Friction is caused under the same contact conditions that are basic to the wear process, however, they may act differently in different situations. The friction and wear of metals are primarily influenced by the interaction between the microscopic peaks and valleys on the surfaces that are sliding against each other. With sliding distance, the condition of mating surfaces changes which influence the friction and wear properties. In Fig. 1.5, the graph illustrates the variation of the COF over time or sliding distance. Friction coefficient has four different stages. During stage I, initially, a low rate of COF depends on low loads and shear resistance of contaminants present on the surface; however, it is independent of a combination of mating materials. The friction coefficient gradually increases with time due to increased adhesion and asperity interactions resulting due to removal of the surface layer.

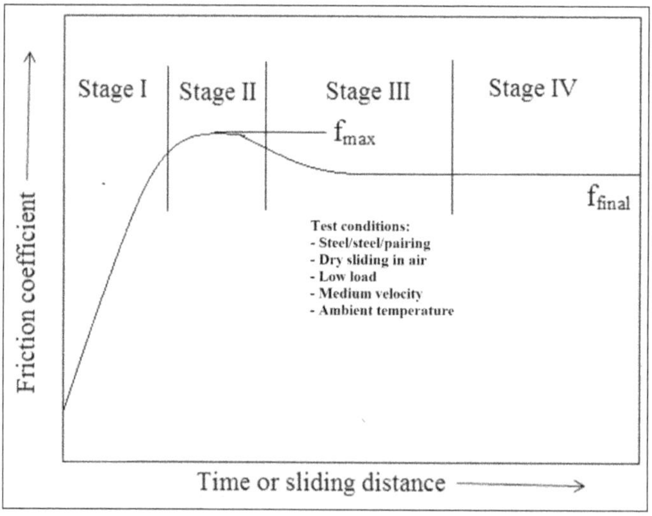

Fig. 1.5 Friction - time curve [43]

Entrapped wear particles also contribute to increased friction coefficient which attains its maximum value during stage II. In stage III friction coefficient decreases due to the formation of a protective tribo-chemical surface layer which reduces the ploughing and asperity deformation. In stage IV steady-state condition exists which leads to a constant value of COF [43].

1.7 FACTORS AFFECTING WEAR AND FRICTION

The wear and friction characteristics are contingent upon a range of structural and operating parameters. Structural parameters include factors such as the materials involved in contact, the choice of lubricants, the environmental conditions, the microstructure, as well as the mechanical and physical characteristics. On the other hand, operating parameters, includes the nature of the contact (whether it involves sliding, rolling, reciprocating, or impact), the magnitude of applied load, the velocity of motion, the temperature of operation and the duration of exposure, also exert significant influence on both wear and friction [58].

1.8 DESIGN OF EXPERIMENTS

The Design of Experiments (DoE) enables the user to acquire more information with less effort. DoE is a systematic and statistical approach that strategically plans and organizes experiments to extract maximum insights while minimizing the number of trials or experiments conducted. It involves manipulating one or more independent factors to examine their effects on one or more dependent variables (responses). The DoE is used when multiple factors are suspected to influence an output. Due to the prohibitive costs associated with physical testing, these methodologies are commonly used in many industries. [59].

1.9 RESPONSE SURFACE METHODOLOGY (RSM)

Response surface methodology (RSM) is a collection of mathematical and statistical methodologies. RSM is used to model and analyse complex relationships between multiple independent variables (factors) and their effects on the dependent variable. In most practical cases, RSM applications entail multiple responses. The input variables are referred to as independent variables and these variables are under the direction of the operator, particularly for the purposes of testing or experimentation. It is widely used for product design, process optimization, quality improvement and robust parameter estimation. [60].

The initial step in RSM involves identifying an appropriate relationship between the response and the factor variables. This requires conducting a well-designed experiment and using the resulting data to create response surface models.

Subsequently, the response surface is examined to determine a direction that leads towards the approximate location of the optimum. Once the optimal region is identified, a more detailed model is developed, typically accompanied by a new set of experiments within this optimal region, to accurately pinpoint the optimum. Hence, RSM is recognized as a sequential process. In RSM, the surfaces are usually modelled by either first-order or second-order models [61]. The types of DoE in RSM methods are depicted in Fig. 1.6.

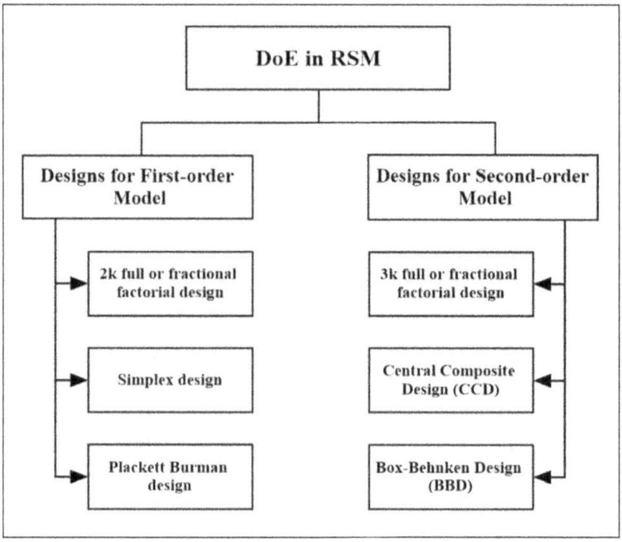

Fig. 1.6. Types of design of experiments in RSM

The Box-Behnken design (BBD) method, derived from statistical and mathematical principles, provides an analytical framework to understand the impact of independent variables on a dependent response. Commonly used for assessing structural responses, this approach treats inputs as variables and interprets the resulting outputs as the key feedback, affected by numerous variables. This method delineates three tiers for every factor with configurations of -1, 0, +1, each comprising a specific subset of the factorial combinations from the 3^k factorial design. The BBD method, within the framework of RSM, is employed to establish a relationship among process parameters. Through the formulation of a mathematical model, the integration of outcomes and their corresponding variables is achieved, enabling the enhancement and optimization of the process. Eq. (1.5) expresses the count of experiments conducted in the Box-Behnken design method.

$$q = 2p(p-1) + C \tag{1.5}$$

When conducting a total of q experiments with p desired parameters and C central points. Fig. 1.7 provides visual representations of three-level factorial designs. The BBD method emerges as a particularly efficient response surface method in comparison to a central composite design method. It proves advantageous by steering clear of extreme factor combinations, enabling users to sidestep the challenges associated with extreme factor levels [62, 63].

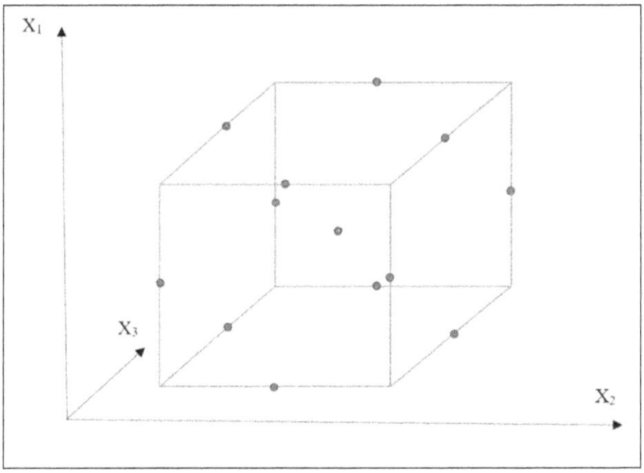

Fig. 1.7 Geometrical view of three-factor BBD

1.10 ORGANIZATION OF THE THESIS

The current dissertation comprises six chapters and a summary of each chapter is outlined below:

Chapter 1 gives a comprehensive overview of composite materials, including their various types and the advantages they offer compared to monolithic materials. Furthermore, it examines MMCs, highlighting their superior attributes relative to other composite varieties. This section delivers a succinct overview of the diverse fabrication techniques employed in the production of particle-reinforced MMCs. Additionally, it offers insights into different categories and phases of wear and friction. Furthermore, this chapter presents an overview of RSM based BBD method.

Chapter 2 provides a comprehensive literature review, focusing on the synthesis, characterization and testing of AMCs produced through various methods. The conventional results from previous studies on the advancement of diverse reinforcements, such as red mud and sillimanite, are explained in detail. Different reinforcements are investigated in AMCs for dry sliding wear and erosive wear. The various DoE methods used in dry sliding wear and air jet erosion are also investigated. After summarizing the literature, a research gap is identified and objectives are outlined.

Chapter 3 gives detailed information about the materials, fabrication methodology and various equipment used in the fabrication, testing and characterization of AMCs and LM25 alloy. This Chapter also discusses the BBD experimental design of input process parameters of dry sliding wear and air jet erosion test.

Chapter 4 discusses the physical, microstructural and mechanical properties of the LM25 alloy and AMCs. The physical analysis contains density, porosity and thermal conductivity analysis. Microstructural characterization includes optical and Scanning electron microscopy (SEM) micrographs of AMCs and the LM25 alloy, along with energy dispersive spectroscopy (EDS) analysis. Tensile strength, fractography and microhardness are included in mechanical properties analysis.

Chapter 5 presents the results of dry sliding wear, friction and erosive wear analysis of AMCs. The optimized dry sliding wear, friction and erosive wear input process parameters obtained from the BBD are also discussed and compared in detail in this section. An analysis of the wear tracks of the worn-out wear specimen is discussed as a final step.

Chapter 6 presents the outline of the key outcomes and notable conclusions that are derived from the experimental work conducted in this thesis. This chapter also provides the primary conclusions to be inferred from the current research work and potential avenues for extending the existing work.

CHAPTER 2

LITERATURE REVIEW

2.1 OVERVIEW

Currently, researchers worldwide are primarily concentrating on aluminium due to its exceptional blend of corrosion resistance, low density and outstanding mechanical characteristics. The distinctive thermal properties of AMCs, including thermal conductivity and the ability to tailor the coefficient of expansion down to zero, further enhance their potential in aerospace, automobile and avionics. In Metal Matrix Composites (MMCs), aluminum is often chosen as the matrix material due to its low density, excellent processing capabilities and ease of manufacturing. AMCs are produced by embedding ceramic particles such as oxides, borides, carbides and nitrides into the aluminum matrix. This reinforcement enhances the material's wear resistance and thermal stability. However, synthetic ceramics can be quite costly, driving up the overall price of the composites. To address this issue, researchers have developed composites that incorporate naturally occurring minerals like sillimanite, garnet, zircon and rutile, as well as industrial by-products such as red mud and fly ash. A hybrid metal matrix composite is created by combining at least two different reinforcing materials, leveraging the benefits of multiple reinforcement components to achieve superior properties compared to composites with a single reinforcement. [3-5].

2.2 ALUMINIUM SILICON ALLOY

Aluminium-silicon (Al-Si) alloys find widespread use across various engineering applications because of their characteristics including low density, high specific strength, excellent castability, high resistance to corrosion and favorable wear properties [64, 65]. Typically, Al-Si alloys with silicon concentrations ranging from 4 to 20 wt.% are utilized in diverse engineering contexts. These alloys are categorized based on silicon content into three main segments: hypoeutectic Al-Si alloys (Si<12.6 wt.%), eutectic Al-Si alloys (Si =12.6 wt.%) and hypereutectic Al-Si alloys (Si>12.6 wt.%). The aluminium-silicon phase diagram depicted in Fig. 2.1 which illustrates this classification. Notably, the only invariant reaction observed is the eutectic reaction, as described in Eq. (2.1).

$$L \rightarrow \alpha + \beta \tag{2.1}$$

Where,
L: liquid phase
α: predominantly aluminium phase
β: predominantly silicon phase

The eutectic transformation of Al-Si alloys occurs at a temperature of 577°C, resulting in a liquid solution phase with a silicon content of 12.6 wt.%. Hypoeutectic Al-Si alloys consist of a combination of proeutectic or primary aluminium phase and an Al-Si eutectic blend. Conversely, hypereutectic Al-Si alloys feature a proeutectic or primary silicon phase alongside the Al-Si eutectic mixture. Typically, hypereutectic Al-Si alloys find use in various engineering applications due to the presence of the primary silicon phase. The presence of hypereutectic silicon in Al-Si alloys contributes to favorable properties such as enhanced castability, hardness and wear resistance [66, 67].

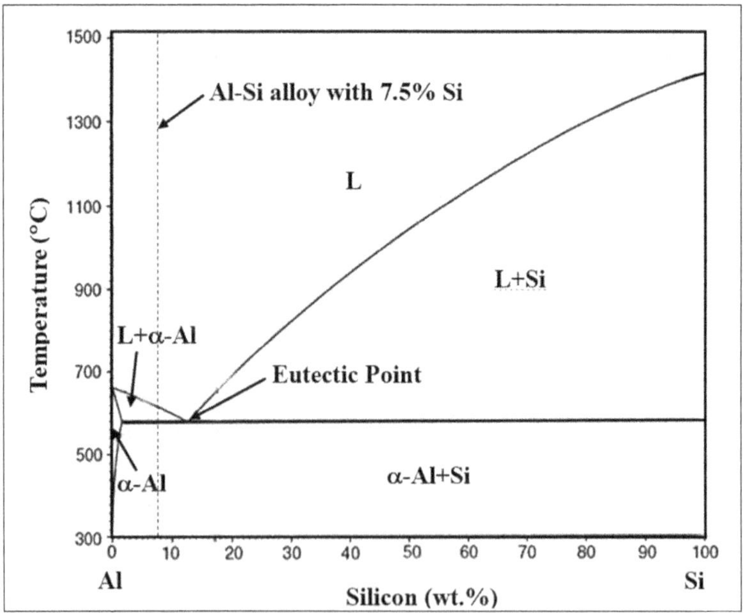

Fig. 2.1 Aluminium-silicon phase diagram

2.3 SILLIMANITE REINFORCED AMCS

Singh et al. [68] have investigated the effect of sillimanite particle reinforcement on LM6 aluminium alloy. A strong mechanical bond was observed between the sillimanite particles and the base alloy. The reinforcing material was evenly distributed throughout the matrix. Although the composite exhibited slightly lower strength compared to the base alloy, it demonstrated significantly higher hardness.

Pethuraj et al. [69] employed the stir casting process to cast AMCs using sillimanite (5 and 10 wt.%) as reinforcement. It was found that incorporating 5 wt.% sillimanite reinforcement improves the tensile strength, impact strength and hardness. Including more than 5 wt.% of sillimanite reinforcement did not notably impact the mechanical properties of the AMCs.

Gupta et al. [70] prepared sillimanite and rutile particle reinforced LM27 aluminium alloy AMCs. Particles ranging from 1 to 32 μm were incorporated at 5, 10 and 15 wt.%. Microstructural examinations demonstrated a homogeneous dispersion of mineral particles within the base alloy. The extension of reinforcement led to the refinement of the Si phase in the LM27 alloy. Rutile-reinforced composites exhibited greater refinement compared to their sillimanite-reinforced counterparts. The hardness of sillimanite-reinforced composites across different phase regions exceeded that of rutile-reinforced composites. The notably elevated hardness value at the sillimanite and base alloy junction indicated a robust interfacial bonding between reinforcement and matrix in the processed AMCs.

Sharma et. al. [71] manufactured AMCs reinforced with sillimanite particles using the stir casting method, incorporating three different sizes of reinforcement particles. The level of reinforcement was adjusted within the range of 3 to 15 wt.%. The optical micrographs exhibited a consistent dispersion of strengthened particles within the processed AMCs. Additionally, the elevated values of the Nano-hardness test are observed at the sillimanite and base alloy interface. It indicates a strong interfacial attachment between the matrix material and sillimanite particles.

Gupta et al. [72] prepared hybrid AMCs using LM27 aluminium alloy reinforced with rutile and sillimanite in 1:1 proportion. Stir casting was used to process hybrid AMCs. Mineral reinforcement particles were selected to be between 1 and 32 μm. The reinforcement was evenly distributed up to 15 wt.%. The structure, arrangement and size of the silicon phase within the matrix material underwent substantial alterations after the inclusion of reinforcement to the base alloy. Consequently, after rutile was added, the silicon phase became refined, whereas sillimanite reinforcement strengthened the interfacial bonding between LM27 alloy and reinforcement. As the amount of sillimanite particles in matrix material increased, hardness increased.

Singhal et al. [73] employed the stir casting technique to prepare the AMCs of LM30 alloy with 10 wt.% sillimanite minerals. Additionally, 1 wt.% of tin and graphite particles were incorporated to create novel composite materials. Micrographs of the AMCs revealed uniform dispersion of graphite, tin and sillimanite minerals. The incorporation of reinforcement material into the base alloy resulted in the enhancement of the silicon morphology. Microhardness analysis indicate a good interfacial bonding between the reinforcement and the base alloy. Furthermore, Brinell hardness test highlighted an improvement in composite hardness with the addition of reinforcements, with the most substantial enhancement observed in composites containing both tin and graphite.

2.4 RED MUD REINFORCED AMCS

Prasad et al. [74] studied the impact of red mud particles on the mechanical behavior of pure aluminium (IE-07 grade). AMCs with red mud are fabricated using the stir casting technique at 10, 15, 20 and 30 wt.% of red mud. AMCs show a homogeneous dispersion of red mud particles in the base alloy. By incorporating red mud particles into the matrix material, the composites become harder.

Singla et al. [75] casted MMCs of Al 6061 reinforced distinctly with alumina (Al_2O_3), red mud and silicon carbide (SiC). The wt.% of SiC, Al_2O_3 and red mud reinforcements are taken at levels of 2.5%, 5%, 7.5% and 10% by weight. A composite having 7.5 wt.% of red mud exhibited the highest microhardness, followed by AMC with 7.5 wt.% of SiC and Al_2O_3.

Kumar et al. [76] fabricated a hybrid AMC comprising Al6061, alumina, red mud and graphite by stir casting route. The variation in wt.% of red mud particles was taken at 3, 7 and 11. The hardness of 11 wt.% red mud composite is 12.44 % higher than that of matrix material, but 21.63 wt.% lower than that of Al6061/graphite (3 wt.%)/ alumina (3 wt.%) composite.

Lokesh et al. [77] successfully utilized the stir casting technique to create AMCs, incorporating inclusions of red mud wt.% of 2%, 4%, 6% and 8%. A homogeneous dispersion of red mud particles in the Al-6061 alloy was accomplished. The examination reveals that the composite containing 2 wt.% red mud exhibits outstanding resistance to indentation, showcasing a 41% rise in hardness in contrast to the sample reinforced with 8 wt.% red mud particles.

Samal et al. [78] examined the mechanical and microstructural characteristics of AA6082 AMCs using red mud as reinforcement. Utilizing a stir casting technique, four distinct metal matrix composites of aluminium were fabricated, featuring varying volume percentages of 0, 2, 4 and 6%. Increasing the quantity of red mud in matrix material improved the tensile strength and hardness. SEM of the fractured tensile samples unveils ductile fracture features in the shape of pits, cracks and micro voids.

Kar et al. [79] fabricated red mud and titanium carbide (3 to 12 % by weight) reinforced aluminium Al-7075 alloy AMCs by stir casting method. There is a 62% and 6% increase in hardness for AMCs reinforced with red mud and TiC, respectively. The impact strength and elongation are reduced in both AMCs.

Hugar et al. [80] investigated the mechanical behaviour of red mud/titanium oxide (TiO_2)/Al6061 aluminium alloy MMCs fabricated by the stir casting method. The Al6061 alloy is reinforced with titanium dioxide in weight proportions of 2%, 4% and 6%, while maintaining a constant 2% of red mud. The findings reveal that the addition of titanium dioxide and red mud enhances the tensile strength, compressive strength and hardness of the AMCs. The ultimate tensile strength and yield strength increase by 29.77% to 72.36% and 29.96% to 63.90%, respectively, compared to the base alloy. Additionally, the hardness improves by 9.41% to 41.17% and the compressive strength increases by 9.28% to 33.20%, relative to the base alloy. Scanning Electron Microscopy (SEM) analysis indicates a uniform distribution of the constituents, ensuring consistent load transfer.

2.5 DRY SLIDING WEAR BEHAVIOUR OF AMCS

Kaushik et al. [81] studied the tribological behaviour of Al-6082/Graphite/SiC hybrid AMCs, comparing them with the matrix alloy. The dry sliding wear tests were done utilizing a pin-on-disc setup under test parameters comprising a 50–75 m sliding distance 5–15 N load and 200 mm particle size. Comparatively to both the base alloy and composite reinforced with SiC, the hybrid AMC demonstrated enhanced wear resistance. A composite Al-6082/Graphite/SiC material exhibited 16.4% greater wear enhancement than a matrix material under increased loads and extended sliding distances.

Baradeswaran et al. [82] studied the effect of graphite on the tribological behaviour of hybrid AMCs composed of Al 7075, Al_2O_3 and graphite. The hybrid AMCs were fabricated with a 5 wt.% addition of graphite particles and different amounts of Al_2O_3 (2, 4, 6 and 8 wt.%). The hybrid AMCs, incorporating graphite, exhibited enhanced wear resistance compared to the matrix material.

Ramkumar et al. [83] utilized the stir casting process to fabricate MMCs of AA7075 alloy reinforced with 2.5 wt.%, 5 wt.% and 7.5 wt.% titanium carbide (TiC). The SEM analysis reveal a homogeneous dispersion of reinforcement in the base alloy. The increase of TiC reinforcement resulted in a notable improvement in the wear resistance of the AMCs.

Liu et al. [84] conducted the synthesis of Al7075 aluminium alloy, incorporating boron carbide (B_4C) as a reinforcement and molybdenum disulfide (MoS_2) as a lubricant, across different amounts (4%, 8% and 12% by weight) employing the stir casting method. The experimental results indicated a uniform dispersion of the reinforcement material throughout the base alloy, presenting fine dendritic structures within the solid aluminium solutions. Remarkably, the incorporation of MoS_2 together with B_4C resulted in a notable improvement in both the wear resistance and COF within the hybrid AMCs.

Suresh et al. [85] employed the stir casting method to fabricate MMCs, introducing varying wt.% (1.0, 2.0, 3.0 and 4.0) of nano-SiC powder as reinforcement in aluminium 7075 alloy. Tribological investigations were carried out through systematic adjustments in the applied load utilizing a pin-on-disc setup to assess the wear rate of the AA7075/nano-SiC AMCs. The results reveal that the insertion of nano-silicon carbide (SiC) powder significantly improve the wear resistance of the AMCs.

Sharma et al. [86] studied the wear characteristics of the Al-2024 alloy and red mud MMCs in a utilizing a pin-on-disc setup. The study investigated the wear behavior of composites under different conditions, such as reinforcement, applied weight, sliding distance and sliding speed. The findings indicated that sliding speed emerged as the highest influential aspect affecting the AMC's wear behaviour, followed by applied weight and red mud amount. Upon examination of the worn surfaces using SEM, it was observed that increasing the reinforcement amount decreased the wear rate.

Geetha et al. [87] explored the influence of different particle sizes of red mud on the characteristics of dry sliding wear and abrasive wear of A356 alloy/ red mud AMCs. AMCs were produced through the stir-casting method by integrating a 15% volume of reinforcement using three different particle dimensions (53, 75 and 103μm). The addition of finer-sized particles led to a reduction in wear rates. The wear examinations additionally unveiled that the abrasive wear rate surpassed the rate of sliding wear. In conclusion, it is affirmed that particle size significantly influences the mechanical properties and wear behaviour of the produced composites.

Samal et al. [88] examined the tribological properties of Al-6082 alloy MMCs reinforced with red mud particles (2 wt.% to 6 wt.%) through the stir-casting technique. Reinforcement particles are homogeneously distributed throughout the base alloy, resulting in good wettability between the base alloy and reinforcement. Wear analysis shows that the wear rate decreases as the amount of red mud increases, but it also decreases with longer sliding distances.

Singh et al. [89] examined the two-body abrasive wear characteristics of a composite of LM6 aluminium alloy with 10 wt.% sillimanite particles. The research encompassed changing applied pressures and abrasive dimensions over various sliding distances. The test outcomes show an increase in the wear resistance as the sliding distance elevated, eventually stabilizing at a constant value. Additionally, an escalation in abrasive size and applied load resulted in an increased wear rate, irrespective of the reinforcement percentage. Notably, at an abrasive size of 25 μm, the composite demonstrated maximum wear resistance compared to the base alloy. However, with a 200 μm abrasive size, the composite demonstrated diminished wear resistance in comparison to matrix material, regardless of the load. The AMC exhibited enhanced wear resistance at an abrasive size of 100 μm compared to the base alloy under lower loads. However, this pattern reversed under higher applied loads.

Singh et al. [90] examined the impact of sillimanite particles on the wear resistance of LM6 alloy MMCs. The AMCs were manufactured through the stir casting route, with 10 wt.% sillimanite reinforcement. Dry sliding wear tests were performed at pressures ranging from 0.2 to 1.6 MPa and velocities of 1.89 to 5.55 m/s. Findings indicated that the AMCs demonstrated a reduced rate of wear in comparison to the base alloy. An escalation in the applied pressure resulted in an elevated wear rate, whereas an augmentation in speed had the converse effect. Moreover, the temperature increases near the contact surface and the COF is lower in the composite than in the matrix material.

Sharma et al. [91] conducted a thorough investigation into the impact of two different size reinforcement, on the wear resistance of LM30/sillimanite AMCs. Coarse and fine particles of sillimanite are taken in weight ratios of 1:3, 1:1 and 3:1. Remarkably, the wear rate of AMC with a 15 wt.% dual particle size and a weight ratio of 3:1 exhibited a notable enhancement of 59% over the matrix material. Upon closer examination through SEM analysis, it was noticed that at lower pressure, the dominant wear mechanism involved a blend of abrasive and adhesive wear. However, as the contact pressure increased, a shift towards delamination wear became more prominent.

Raju et al. [92] examined the tribological characteristics of Al-1100 alloy reinforced with coconut shell ash (CSA) AMCs at high temperatures which was fabricated by stir casting method. The wear rate of AMCs improved in comparison with base alloy for the rise in applied pressure at elevated temperatures. A marginal alteration in coefficient of friction (COF) values was noted as the contact pressure increased under room temperature conditions. At higher temperatures, the COF increased till an applied pressure of 6 MPa, after that the formation of oxides reduced the COF of AMCs.

Sharma et al. [93] studied the wear resistance of LM30/sillimanite AMCs at elevated temperatures. Sillimanite particles, ranging in size from 1 to 20 μm and 75 to 106 μm, were combined in weight proportions of 3:1, 1:1 and 1:3. The composite with 15 wt.% sillimanite and a particle ratio of 3:1 (fine to coarse) exhibited the highest wear resistance at elevated temperatures. At 200-degree Celsius temperature and 1 MPa pressure, the 15 wt.% sillimanite and a particle ratio of 3:1 (fine to coarse) composite demonstrated superior mechanical properties, including a 15% higher hardness, an 80% lesser wear rate, a 48% lower COF and a 28% coefficient of thermal expansion compared to the matrix material. SEM test of the wear out paths and wear debris revealed the distinct wear mechanisms accountable for material loss at distinct applied pressures.

2.6 EROSIVE WEAR BEHAVIOUR OF AMCS

Khan et al. [94] investigated the erosive wear rate of an AMC reinforced with 10 wt.% Silicon Carbide (SiC). The AMCs were fabricated by incorporating silicon carbide particles having particle size ranges of 50-100 μm into the matrix alloy. The wear response was analysed using the sample rotation technique with a slurry pot erosion tester. The investigation explored the impact of velocity and sand concentration on the wear rate. Microstructural examinations revealed a robust interfacial bonding among the matrix material and the SiC particles, accompanied by a homogenous dispersion of reinforcement particles. In acidic and saline environments, the composite demonstrated enhanced wear resistance. The results show a relation between increased sand concentration and a high rate of material loss, attributed to the intensified impinging action of the sand particles. Moreover, it was found that the erosion rate exhibited a combination of influences with changes in rotational speed.

Bera and Acharya [95] studied the erosive wear behaviour of LM6/cenosphere composites. Experiments were conducted by varying the reinforcement weight percentages (5 wt.%, 7.5 wt.%, 10 wt.% and 12.5 wt.%) across four different speeds (48 m/s, 70 m/s, 82 m/s and 109 m/s) and impact angles (30°, 45°, 60° and 90°), using silica as the abrasive medium. The SEM analysis suggested that the erosion rate of the AMCs was influenced by both impingement angle and erodent speed. The incorporation of reinforcing elements elevated the erosion rate, particularly under conditions of erodent speed and impingement angle. The identified erosion structure encompassed micro-cutting and ploughing at low impingement angles, whereas composite erosion through fracture occurred at high impact angles.

Longalayuk et al. [96] fabricated Al-7075/SiC MMCs by stir casting route with the incorporation of 5%, 10% and 15% SiC powder along with a 1% addition of Magnesium. The erosion test was conducted to systematically investigate the erosion resistance of the MMC by employing Alumina powder as the abrasive material. The results obtained from the Air Jet Erosion Testing revealed a notable improvement in erosion resistance when 5% SiC powder was added to the base alloy. A further rise in the SiC content to 10% and 15% did not lead to a significant improvement in erosion resistance.

Pethuraj et al. [69] employed vacuum-assisted stir casting to cast MMC reinforced with sillimanite, comprising 5% and 10% wt.%. The erosive wear behavior of AMCs was systematically evaluated under various test parameters, utilizing alumina as the erodent material. The experimental setup involved erodent speeds of 41 m/s, 72 m/s and 100 m/s, with impingement angles of 45°, 60° and 90°. Observations revealed that, at higher speeds, the erosion rate increased, despite of the type of AMCs or matrix material. Additionally, at lower impingement angles, the erosion rate of base alloy surpassed that of the AMCs. The introduction of sillimanite particles decreased the erosion rate, reaching a notable reduction within the range of up to 5 wt.%. However, beyond this reinforcement level, there was no significant impact on the erosion rate. Intriguingly, the erosion rate of the AMC with 10 wt.% reinforcement was found to be higher than that of pure Aluminium.

Bera et al. [97] employed the squeeze casting method to fabricate MMC, utilizing fly ash as reinforcement (2.5, 5, 7.5, 10 wt.%) and A357 aluminium alloy as the matrix material. Additionally, the erosive behavior of AMCs was investigated through solid particle erosion tests conducted at four velocities (48 m/s, 70 m/s, 82 m/s and 109 m/s) and impingement angles (30°, 45°, 60° and 90°) using silica particles as the abrasive material at room temperature. The findings highlighted the significant influence of both impingement angle and erodent speed on the erosive behavior of the AMCs. The rate of erosion demonstrated an escalation at elevated impact velocities, regardless of alterations in impingement angle and the wt.% of reinforcement. The observed erosion mechanisms for the AMCs were identified as micro ploughing and micro-cutting.

2.7 DESIGN OF EXPERIMENTS IN DRY SLIDING WEAR AND AIR EROSION BEHAVIOUR OF AMCS

Kumar et al. [98] studied the tribological behaviour of Al-7075 alloy reinforced with 7 wt.% Silicon carbide and 3 wt.% graphite, prepared by the stir casting process. Wear tests were performed in dry state across different loads (20–60 N), sliding speed (2–6 m/s) and sliding distances (2000–4000 m). Under varying combinations of loads, sliding distance and sliding speed, the hybrid AMC demonstrates a lower specific wear rate than unreinforced Al 7075. Based on RSM, the applied load was determined to be the primary element influencing the specific wear rate, followed by sliding speed and sliding distance.

Rajesh et al. [99] examined the tribological behavior of red mud reinforced MMC under various loading conditions, sliding speeds and red mud wt.%. Specific wear rate and COF were identified as output quality characteristics. In this study, Taguchi-based L9 orthogonal arrays were used for experimentation and ANOVA was used to determine optimal settings. It was noted that the most effective factor configurations differed for each performance output. To concurrently reduce both responses, Multi-Objective Optimization utilizing Ratio Analysis (MOORA) was utilized. MOORA analysis unveiled that the most effective configuration for dry sliding wear conditions in aluminium derived from red mud includes a normal force of 20N, sliding velocity set at 3 m/s, reinforcement percentage at 20% and a material counter face hardness of 58HRC.

Baradeswaran et al. [100] employed the liquid casting technique to reinforce Al-6061 and Al-7075 with 5 wt.% of graphite and 10 wt.% boron carbide (B_4C). Wear trials were executed utilizing a pin-on-disc apparatus, altering factors including applied force (10-30 N) sliding distance (1000-2000 m) and sliding speed (0.6-1.0 m/s). Mathematical models, generated through response surface regression analysis, were developed and contour plots illustrating these models were created for Al-6061, Al-7075 and their composite hybrids. RSM results revealed that the combination of 10 newton force, 0.8 m/s sliding speed and 2 km sliding distance achieves the lowest wear rate.

Sharma et al. [101] examined the influence of graphite particles on the tribological behaviour of AMCs based on the Al6082 alloy. The investigation involved varying the wt.% of graphite from 3% to 12%, with an increment of 3%. The wear rate of the AMCs was estimated by a pin-on-disc machine, where process variables included the proportion of reinforcement, applied load, sliding distance and sliding velocity. RSM was utilized for experiment planning and analysis. The findings highlighted that sliding distance arrives as the highest influential factor, while load has the least impact on wear. The validation test illustrated that the deviation between the experimentally observed and predicted wear rate values is within the margin of 3–10% which validates the accuracy of the model.

Singh et al. [102] examined the wear rate of titanium diboride (3, 6, 9 and 12 wt.%) reinforced Al-6082 alloy composites, prepared by stir casting method. Wear tests were carried out utilizing a pin-on-disc setup under ambient conditions, considering variables such as the proportion of reinforcing material, sliding velocity, applied force and distance of sliding. After RSM analysis results revealed that increased titanium diboride (TiB_2) particle content and sliding velocity led to lower wear rate, while increased applied load and sliding distance resulted in improved wear rate in AMCs. ANOVA determined that the primary influential factor was the sliding distance, accounting for 65.28%, trailed by sliding speed at 14.78%, load at 9.39% and reinforcement percentage at 3.86%. Through confirmation tests, the mathematical model was verified as an accurate predictor of wear rate.

Madhavaraoa et al. [103] explore the dry sliding wear rate of Al-7075 aluminium matrix reinforced with silicon carbide particles (10 and 15 wt.%), produced through the stir casting method. Wear and frictional characteristics of metal matrix composites were examined using a pin-on-disc machine. L-9 Orthogonal array, which is part of Taguchi's method, was used to design the experimental plan. Findings suggest that the distance of sliding exerts the most noteworthy influence, trailed by the load applied and sliding speed. Remarkably, the composite with 10 wt.% is anticipated to achieve the highest precision, reaching 95.2%, while the 15 wt.% AMC is projected to attain a close accuracy of 93.4%.

Kumar and Murugan [104] explored the tribological characteristics of red mud particle-reinforced Al6061/Alumina/Graphite MMC using a Pin on Disc setup. The synthesis of AMCs commenced with the stir casting route, where the wt.% of red mud particles were altered to 3%, 7% and 11%. Wear resistance assessments were carried out using dry sliding tests, employing an L27 orthogonal array. The results demonstrated a notable 90% enhancement in wear resistance for the hybrid composite reinforced with 11wt.% red mud, as compared to the Al6061/Alumina/Graphite Hybrid

MMC. The COF for all composites increased and saturated with applied load and sliding distance, with 11wt.% red mud reinforced hybrid composite showing a maximum 48% increase in COF. Metallographic investigation indicated a shift in wear mechanisms from abrasive to adhesive at the highest value of load, sliding speed and sliding distance for Al6061/Alumina/Graphite Hybrid MMC, while the addition of red mud particles introduced combined abrasive and adhesive wear mechanisms. Grey Relational Analysis (GRA) was utilized for optimizing tribological parameters, uncovering enhanced tribological characteristics in the case of 11wt.% red mud reinforced hybrid composite.

Dey et al. [105] examined the erosion wear behaviour of LM13 alloy with SiC particle reinforcement. The BBD method was employed as the experimental design methodology. Results from the investigation indicated that the incorporation of reinforcement improves the wear resistance of the AMCs. Conversely, contrasting trends were noted in experiments conducted in basic environmental conditions. Additionally, the study identified an optimal processing condition associated with a minimal wear rate threshold. Confirmation tests were carried out to validate these optimal results, including determining the percentage error.

2.8 SUMMARY OF THE LITERATURE

- Among researchers, the most prevalent technique employed for the production of AMCs is stir casting. AMCs can be efficiently and economically manufactured using stir casting. There have been reports of quite good tribological properties achieved by this method.

- Across these studies, sillimanite consistently demonstrated advantageous characteristics, including stable mechanical bonding with the base aluminium alloy, uniform dispersion within the matrix material and a notable refinement of the silicon morphology upon incorporation. Despite a marginally lower strength observed in some instances, sillimanite-reinforced composites exhibited a significantly higher hardness compared to the base alloy. Microhardness studies consistently highlighted a robust interfacial bonding between sillimanite mineral and the matrix material, emphasizing its effectiveness as a reinforcing material. For aluminium matrix composite casting, sillimanite proved to be a versatile and reliable choice.

- Researchers have investigated the impact of incorporating red mud into various aluminium alloys through stir casting techniques, examining the mechanical properties and microstructural behaviour of resulting MMCs. The research findings collectively favour red mud as a viable and beneficial reinforcement material, showcasing its ability to improve the tensile strength, hardness and overall wear behavior of AMCs. Studies show that red mud can enhance the performance of these materials sustainably and effectively.

- Many authors have broadly explored the impact of reinforcement size and reinforcement type on the tribological behavior of AMCs. A subset of studies has specifically investigated the influence of dual reinforced particles on AMC wear characteristics. Incorporating ceramic particles, like red mud, sillimanite, fly ash, SiC, Al_2O_3 and B_4C, has been found to enhance both mechanical characteristics and tribological behaviour due to the high hardness of these additives. Additionally, improved bonding and microstructure refinement result from the disparate coefficients of thermal expansion between the ceramic particles and matrix material. As the concentration of ceramic minerals as reinforcement increases, there is a notable enhancement in wear characteristics up to a specific threshold. In the course of wear testing, the formation of oxides and the development of a mechanically mixed layer (MML) play pivotal roles in mitigating wear in composite materials. As a result of this comprehensive approach, ceramic reinforcements in AMCs are highlighted as having a wide range of advantages.

- Many authors applied different Design of Experiment methods to design their experiments. These methods included the various types of RSM. Reducing physical experimentation, the Box-Behnken Design (RSM) was shown to be more effective in establishing a correlation between processing variables and significantly impacting performance metrics. ANOVA (Analysis of Variance) stands out as a powerful statistical tool, ideal for optimization across various disciplines, particularly when working with experimental designs involving several factors. It is particularly effective for identifying the influence of different factors and their interactions on a particular outcome, making it ideal for optimization studies.

2.9 OBJECTIVES OF THE CURRENT INVESTIGATION

The current investigation aims to synthesize a substitute material for automotive applications with improved mechanical, thermal and tribological behaviour. To attain the objective, the subsequent steps are followed:

- Synthesis of LM25 alloy/ sillimanite/ red mud AMCs with varied wt. % of red mud and sillimanite particles by stir casting process.

- To conduct a microstructural characterization to thoroughly examine the morphology, size, distribution and interfacial characteristics of reinforcement particles within the LM25 alloy. To perform an EDS analysis to confirm the existence of both reinforcement particles in the matrix.

- To examine the impact of reinforcing particles on the physical, microstructural and mechanical properties of LM25 alloy/ sillimanite/ red mud composites at ambient temperature.

- Regression modelling and optimization of dry sliding wear parameters like sliding velocity, applied load and reinforcement wt.% on wear rate and friction characteristics of LM25 alloy/red mud/sillimanite composites at ambient temperature using Box-Behnken Design method.

- Regression modelling and optimization of erosive wear parameters like impingement angle, erodent velocity and reinforcement wt.% on erosive wear rate characteristics of LM25 alloy/red mud/sillimanite composites at ambient temperature using Box-Behnken Design method.

- Study of worn and eroded surfaces under SEM to understand the various wear mechanisms involved during the sliding wear and erosive wear processes.

CHAPTER 3

EXPERIMENTAL DETAILS

3.1 INTRODUCTION

This chapter outlines the details of the reinforcing materials, the synthesis procedure for the composites and the various equipment used for characterizing the composites. The experimental investigation unfolded in three stages. In the initial stage, the process and equipment used in the fabrication, physical and mechanical testing of AMCs are discussed. The second stage discusses the testing procedures and equipment setup for evaluating dry sliding and erosive wear behaviour. The third stage of the investigation involved the design of multiple parameters of dry sliding wear and erosive behaviour using the Box–Behnken design method.

3.2 SELECTION OF MATERIALS

3.2.1 Matrix Material

LM25 alloy which is the aluminium silicon alloy is used as the matrix material in the current research. Its silicon content is approximately 6.5-7.5 wt.%. The LM25 alloy is generally used when excellent castability is required in castings of a particular shape or dimension. The LM25 alloy includes magnesium, iron, nickel, titanium, zinc, manganese, copper, silicon, tin and aluminium. The alloy has good corrosion resistance and is used in a variety of applications such as automobiles, marine and transport industry. It is commonly utilized for alloy wheel rims, engine heads, engine blocks and various other engine components [106, 107]. In the present study, LM25 alloy was imported from Sachin Steel Centre, Mumbai (Maharashtra). The elemental examination of the LM25 alloy is done by Optical Emission Spectroscopy (OES) as shown in Table 3.1.

Table 3.1 Elemental analysis of LM25 alloy

Element	Wt.%
Si	6.989
Mg	0.347
Fe	0.113
Ni	0.015
Ti	0.010
Zn	0.005
Mn	0.003
Sn	0.013
Cu	0.001
Al	Balance

3.2.2 Reinforcing Materials

Red mud and sillimanite serve as reinforcing agents in this research. Both red mud and sillimanite are readily accessible and cost-effective materials for use as reinforcement in AMCs. India's coastal areas are rich in beaches and minerals, offering a wealth of natural resources. $Al_2O_3SiO_4$, an aluminium silicate present in high-temperature, regionally metamorphosed clay-rich rocks, typically appears as hair brown, greyish brown, greyish green, greyish white, or pale olive green, although it is colourless in its pure state. This mineral has a specific gravity ranging from 3.23 to 3.34 and a Mohs hardness of 6 to 7. In its pure form, sillimanite melts at 1850°C [108]. Sillimanite is not only readily available but also boasts excellent thermal stability, corrosion resistance, hardness and a low coefficient of thermal expansion [109]. The compositional details of sillimanite done by chemical analysis are detailed in Table 3.2.

Table 3.2 Compositional details of sillimanite

Compound	Al_2O_3	SiO_2	TiO_2	Fe_2O_3	CaO	MgO	Others
Wt. %	57.00	34.00	1.5	0.65	0.07	0.16	Balance

Bauxite residue, commonly referred to as red mud, is a by-product of the Bayer process used to extract aluminium from bauxite ore. This residue comprises a mix of solid and semi-solid waste materials with various oxides such as iron oxide, titanium oxide and silicon dioxide. The melting point of red mud is estimated to range from 1200 to 1250 °C [110-112]. Red mud is obtained from Nangal Enterprise in Jamnagar,

Gujarat, India, with a particle size of 10-50 µm. Sillimanite is supplied by Vstake Solutions Pvt. Ltd in Navi Mumbai, India, with a particle size of 50-150 µm. Compared to Red Mud, Sillimanite contains more Al_2O_3, while Red Mud has a higher Fe_2O_3 content. Both sillimanite and red mud are utilized as reinforcement materials in the production of hybrid LM25 alloy matrix composites. The compositional details of red mud done by chemical analysis are detailed in Table 3.3. The photograph and SEM images of both reinforcement powders are illustrated in Fig. 3.1a-d.

Table 3.3 Compositional details of red mud

Compound	Al_2O_3	SiO_2	Na_2O_3	TiO_2	Fe_2O_3	CaO	Others
Wt. %	20.20	13.50	9.40	4.00	35.04	5.3	Balance

Fig. 3.1 Showing (a and c) photograph and SEM image of red mud powder respectively (b and d) photograph and SEM image of sillimanite powder respectively

3.3 EXPERIMENTAL SETUP OF THE SYNTHESIS PROCESS

The majority of goods for commercial use are produced through liquid state methods. It was decided to fabricate LM25 alloy composites reinforced with red mud and sillimanite particulate by using the stir casting process. The stir casting setup used for the purpose is shown in Fig. 3.2. The microprocessor-based muffle furnace with a bottom pouring facility is used for the preparation of the LM25 alloy/red mud/sillimanite particulate composite. It is an electrically heated single-phase resistance furnace operating at 220 volts capacity fitted with 8 SiC heating rods. The electrical resistance type furnace was used in this work with a PID automatic temperature controller. It has a temperature control accuracy of ±2°C and a maximum working temperature of 1,400°C. This stir casting machine uses a four-blade graphite stirrer driven by a 0.5 Horsepower Remi DC motor with a 60° blade angle from the bottom. The maximum stirring speed of an electrical stirrer is 2000 rpm.

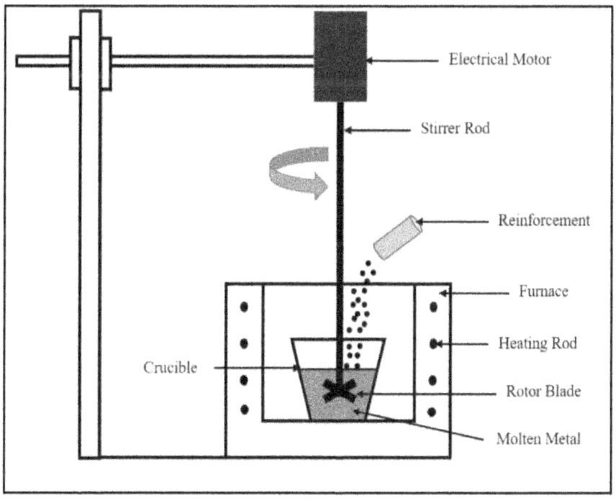

Fig. 3.2 Schematic diagram of stir casting machine

3.4 SYNTHESIS PROCESS OF AMCS

A stir casting method was employed to fabricate the LM25 alloy/sillimanite/red mud metal matrix composite. Three different composites are cast, having 5, 10 and 15 wt.% of sillimanite and red mud in equal amounts. For casting, a graphite crucible was positioned in the muffle furnace with small parts of the LM25 alloy. The furnace is maintained at a 750 °C temperature to melt the LM25 alloy. Stir casting was carried out at 750 °C, which is higher than the melting range of LM25 aluminium alloy (555–620 °C), to ensure sufficient fluidity of the molten metal. Higher fluidity promotes uniform mixing and effective wetting of the red mud and sillimanite reinforcements within the matrix. After heating to melting temperature, the base alloy is stirred for five minutes at 550 rpm. Red mud and sillimanite particles are preheated to 350 °C for 30 minutes in a separate furnace to remove volatile substances. The preheated reinforcement material is then inserted into the crucible and stirred at 250 rpm. After completion of the addition of reinforcing material, the stirrer speed is maintained at 550 rpm for 10 minutes. Fig. 3.3a and b represent the setup of stir casting and the sectional view of the furnace respectively. A cast iron mould (20 x 20 x 150 mm) was filled with the molten mixture and cooled in the air to ambient temperature after the stirring process. In the preparation of the other composites and LM25 alloy, similar procedures were followed and the cast ingots are shown in Fig. 3.4. The cast ingots were termed as SR-5, SR-10 and SR-15 as depicted in Table 3.4.

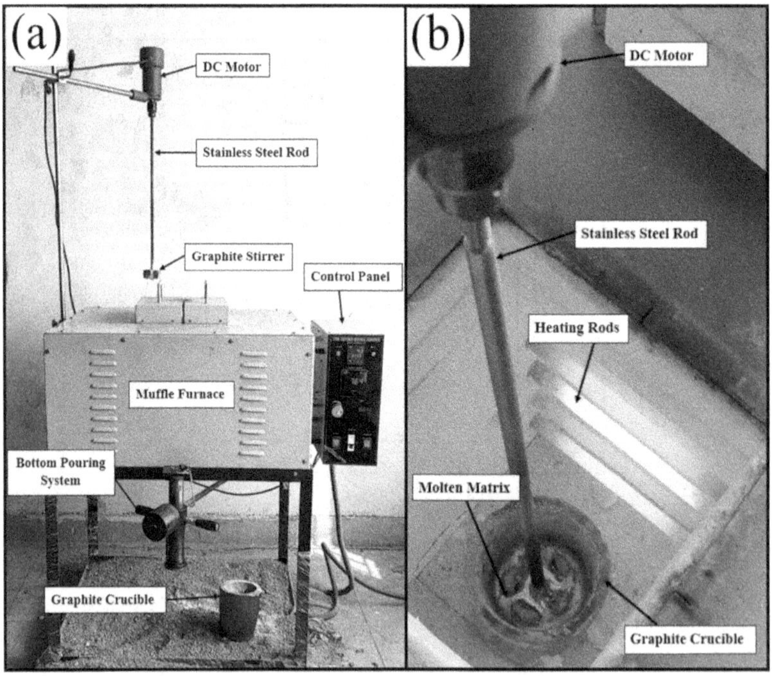

Fig. 3.3 Showing (a) stir casting setup and (b) sectional view of furnace

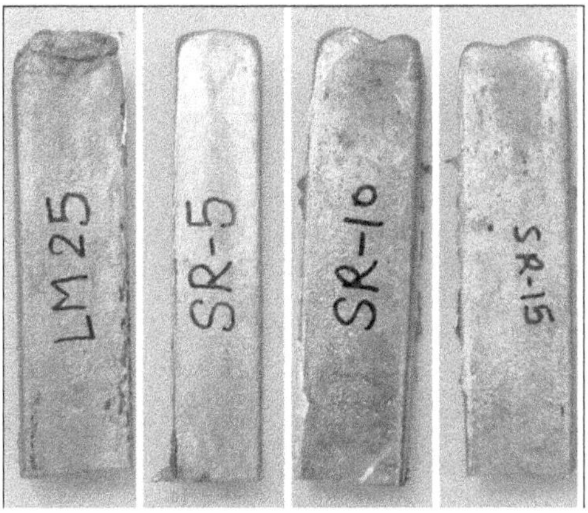

Fig. 3.4 Fabricated ingots

Table 3.4 Nomenclature of aluminium matrix composites

Nomenclature of aluminium matrix composites	Composition of aluminium matrix composites
SR-5	LM25 alloy + 2.5 wt. % Sillimanite + 2.5 wt. % Red mud particles
SR-10	LM25 alloy + 5 wt. % Sillimanite + 5 wt. % Red mud particles
SR-15	LM25 alloy + 7.5 wt. % Sillimanite + 7.5 wt. % Red mud particles

3.5 DENSITY MEASUREMENT

The theoretical densities of the AMCs and alloy were calculated using the rule of mixtures. The theoretical density of the AMCs is calculated by the Eq. (3.1)

$$\rho_T = \rho_M V_M + \rho_R V_R \qquad \textbf{(3.1)}$$

Where ρ_T is the theoretical density of the AMC, ρ_M and ρ_R is the density of LM25 alloy and reinforcement respectively, V_M and V_R is the volume of the LM25 alloy and reinforcement respectively.

The experimental density of AMCs and LM25 alloy is calculated using Archimedes' principle. According to this principle, solid bodies immersed in liquids lose the same amount of weight as the liquid they have displaced. The Archimedes drainage method was used to measure the density of LM25 alloy and AMCs as per the Eq. (3.2).

$$\rho_E = \frac{m}{(m-m_w)}\rho_w \quad \textbf{(3.2)}$$

Where ρ_E is the experimental density of the LM25 alloy or AMC, m_w is the mass of the LM25 alloy or AMCs in distilled water, m is the mass of the LM25 alloy or AMCs in air and ρ_w is the density of distilled water. The density of distilled water at 20°C is 998 kg/m^3 [113].

3.6 POROSITY MEASUREMENT

A significant challenge in composite materials lies in the presence of porosity, which tends to be introduced during the fabrication processes. This porosity can adversely impact the material's inherent characteristics, thereby influencing its overall performance in practical applications. It is crucial to address and minimize porosity, as it can compromise the structural integrity and mechanical properties of the composite. The porosity of the LM25 alloy and AMCs were calculated from the Eq. (3.3).

$$Porosity\ (\%) = \left(1 - \frac{\rho_E}{\rho_T}\right) X\ 100 \quad \textbf{(3.3)}$$

Where ρ_T is the theoretical density and ρ_E is the experimental density of the composite.

3.7 MICROSTRUCTURAL CHARACTERIZATION

To analyse the uniform mixing and morphology of the reinforced material in the base alloy, optical microscopy and scanning electron microscopy are performed. Samples of matrix alloy and composites for microstructures were extracted from cast specimens, then meticulously polished using a sequence of belt grinding and abrasive papers (P220, P400, P600, P800, P1000 and P1200), followed by a final polishing step employing rotating disc cloth. The polishing process setup is illustrated in the Fig. 3.5a and b. As soon as the polishing was complete, the samples were washed thoroughly in running water and dried with hot air blasts. Keller's reagent (95 mL H_2O, 1.5 mL HCL, 2.5 mL HNO_3, 1.0 mL HF) is used as an etching agent.

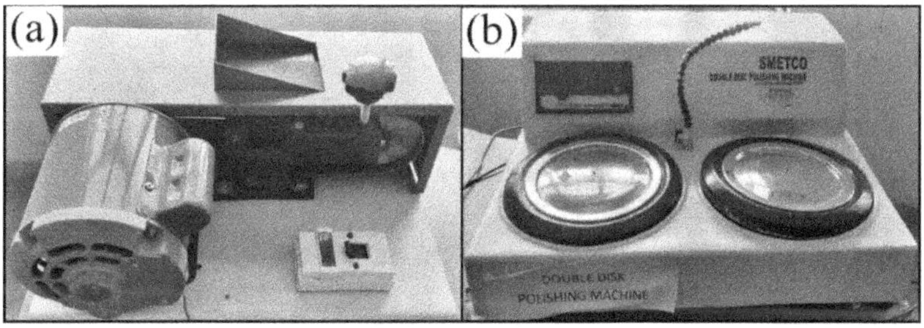

Fig. 3.5 Showing (a) belt grinding machine and (b) double disk polishing machine

Optical microscopy was conducted using a Leica DMI 5000M inverted microscope. The scanning electron microscopy analysis is performed with Zeiss Gemini 300 FE-SEM and energy dispersive spectroscopy analysis is done with Carl Zeiss EVO 50 (WSEM) at different magnifications. An energy dispersive spectroscopy analysis is also done to validate the presence of the reinforcement in the matrix. Fig. 3.6a and b represent the setup of the optical microscope and scanning electron microscope respectively.

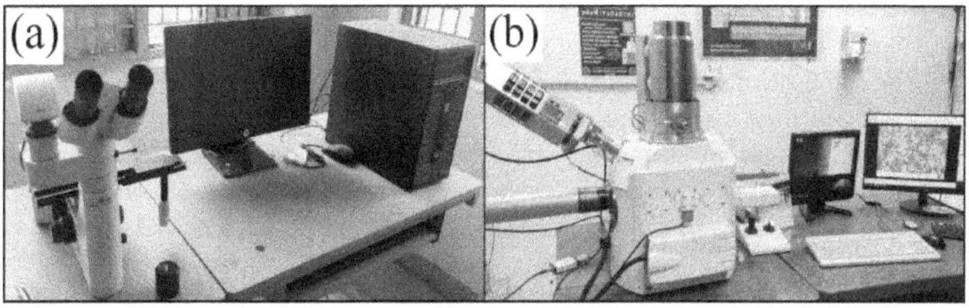

Fig. 3.6 Showing (a) optical microscope and (b) scanning electron microscope

3.8 TENSILE TEST

This study examines the tensile behaviour of AMCs utilizing a universal testing machine (Instron-1195) having 100 kN load capacity as shown in Fig. 3.7a. The specimens utilized in the tensile test had a diameter of 6 mm and a gauge length of 30 mm. A uniaxial load is then applied to both ends of the specimens. Samples of all composites along with base alloy are tested according to the ASTM E8M-16a standards at the strain rate of 0.5 mm/minute [114]. The dimensions of the tensile sample with 2D drawing are shown in Fig. 3.7b and c.

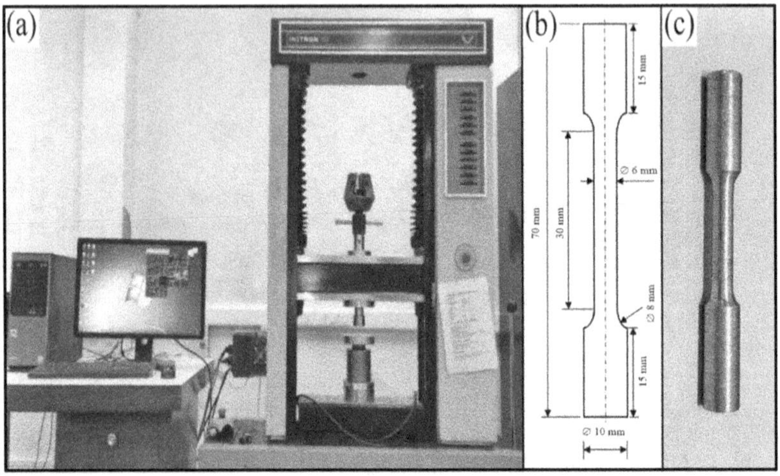

Fig. 3.7 Showing (a) tensile testing machine, (b) tensile specimen 2D drawing and (c) tensile test specimen

3.9 VICKERS MICROHARDNESS TEST

Microhardness measurements were conducted with a Vickers microhardness tester (HDT-B-HV1F). The test force ranges from 10 gf to 1000 gf, with load control set to automatic for loading, dwell and uninstallation phases. The dwell time can be adjusted from 1 to 99 seconds, with increments of 1 second. Microhardness analysis was conducted on aluminium matrix composites according to ASTM E-384 standards [115]. The testing surface underwent preparation to achieve a metallographic polish. The sample is tested under constant 100 gf force and a dwell time of 10 sec. Testing was done at five different points across the sample and average values with standard deviation are reported. Fig. 3.8 illustrates the hardness machine used for testing the hardness of LM25 alloy and AMCs.

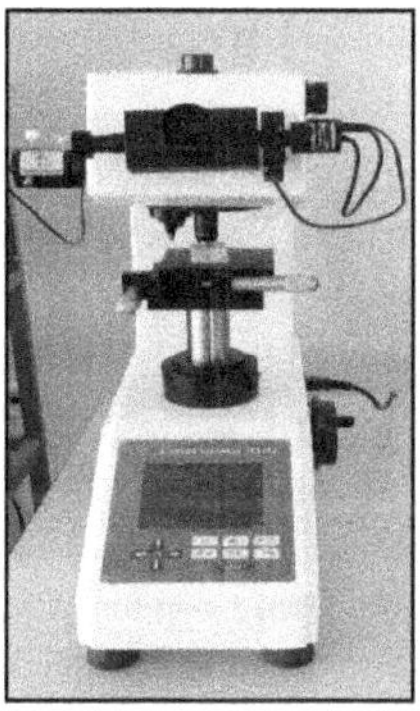

Fig. 3.8 Vickers hardness tester

3.10 THERMAL CONDUCTIVITY TEST

The thermal conductivity of materials is evaluated using the Tci Thermal Conductivity Analyzer by C-Therm as shown in Fig. 3.9a. The instrument has a measurement range of 0-500 W/m-K. Thermal conductivity tests are conducted by ASTM E1225-99 [116]. A cylindrical specimen was used as a reference specimen for the thermal conductivity analyser based on its dimensions. A sample measuring 50 mm in length and 8 mm in diameter has been used as presented in Fig. 3.9b and c.

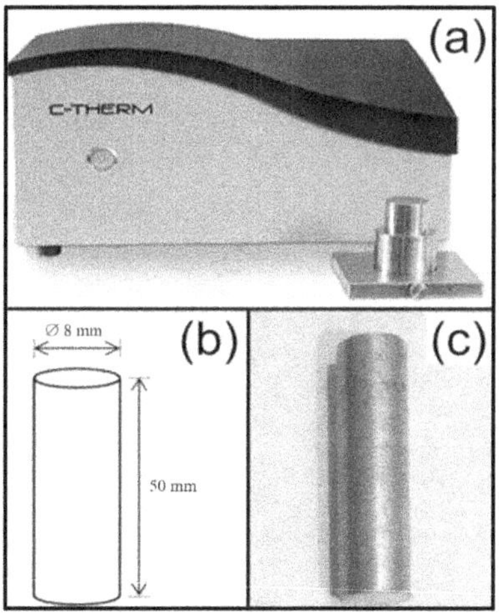

Fig. 3.9 Showing (a) tci thermal conductivity analyser, (b) thermal conductivity specimen 2D drawing and (c) thermal conductivity specimen

3.11 DRY SLIDING WEAR STUDY

The alloy and composite samples were tested for wear resistance using a pin-on-disc setup (Veekay Industries). A wear and friction monitor are presented in Fig. 3.10a. In this setup, the pin was fixed in place while it pressed against the moving surface of a disc with a 100 mm diameter path, made of EN-32 steel with a hardness level of HRC60. The rotating disc with a pin holder is shown in Fig. 3.10b.

According to ASTM G99-05, the wear test samples are 8 mm in diameter and 30 mm in length, as shown in Fig.3.10c [117]. Three variables (reinforcement wt.%, normal load and sliding speed) are taken for the experiments. Each experiment has a fix sliding distance of 3000 m. Before conducting the tests, the test samples underwent metallographic polishing using emery papers. Subsequently, they were meticulously cleaned in acetone, ensuring the removal of any impurities. Following the cleaning process, the samples were carefully dried and weighed utilizing a high-precision electronic balance, as depicted in Fig.3.11. The electronic balance employed for this purpose had a resolution of 0.0001 grams.

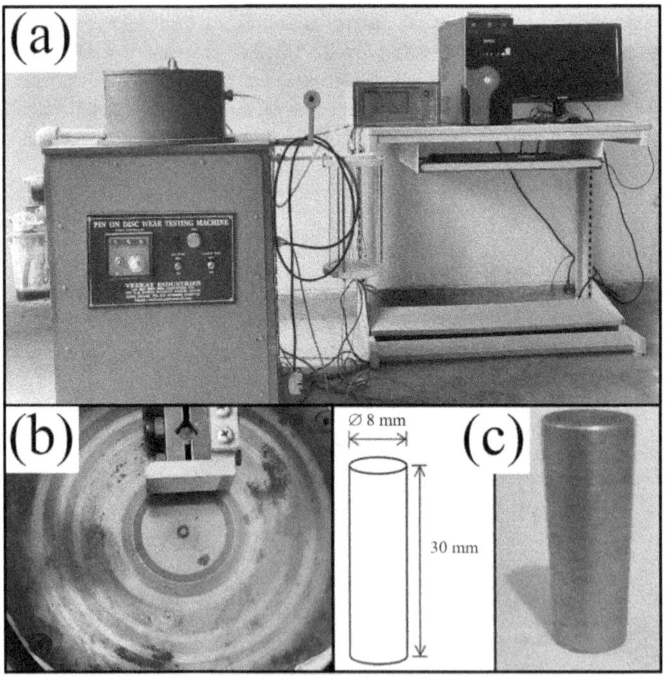

Fig. 3.10 Showing (a) pin-on-disc wear testing machine, (b) rotating disc and (c) pin specimen with 2D drawing

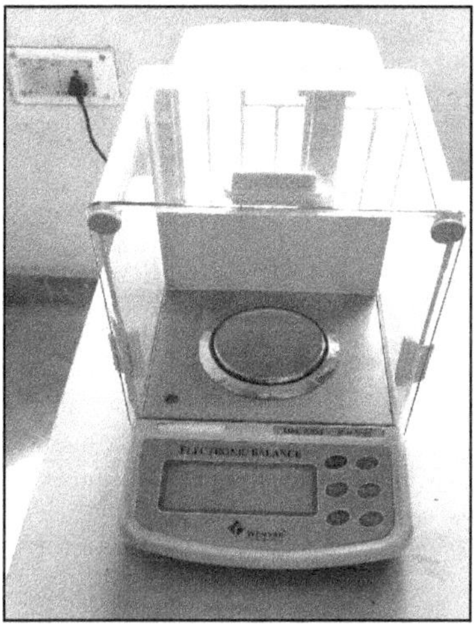

Fig. 3.11 Electronic balance

The following terms were studied in the dry sliding wear performance:

- Mass loss (Δm) is defined as the weight loss of the specimen. It is calculated before and after the run using Eq. (3.4).

$$\Delta m = (m_1 - m_2)\ gram \tag{3.4}$$

Here, m_1 represents the mass of the specimen before the wear test and m_2 indicates the mass of the sample after the wear test.

- In order to calculate volume loss (ΔV) relating mass loss (Δm) to experimental density (ρ_E), Eq. (3.5) is used.

$$\Delta V = \frac{\Delta m}{\rho_E}\ m^3 \tag{3.5}$$

- A wear rate (W_r) is calculated by relating volume loss (ΔV) to sliding distance (SD) as shown in Eq. (3.6).

$$W_r = \frac{\Delta V}{SD}\ mm^3/m \tag{3.6}$$

- Each wear test is averaged over the total number of passes to determine the frictional force developed during the experiment. The Eq. (3.7) is used to determine the average value of the coefficient of friction (μ).

$$\mu = \frac{F_f}{F_n} \qquad \textbf{(3.7)}$$

Where F_f is the frictional force between the disc and the pin and F_n is the normal force on the pin.

3.12 REGRESSION MODELLING AND OPTIMIZATION OF DRY SLIDING WEAR AND FRICTION PARAMETERS

The Box-Behnken Design (BBD) was employed to assess the individual and interactive effects of reinforcement weight fraction (R), normal load (L) and sliding speed (S) as independent variables on the weight loss of the AMCs. Table 3.5 presents the different levels of the parameters. Each parameter was categorized with coding values of -1 for low, 0 for moderate and +1 for high.

Table 3.5 Parameters considered in dry sliding wear test process and their levels

Process Parameter	Notation	Unit	Levels		
			-1	0	+1
Reinforcement	R	Weight %	5	10	15
Load	L	Newton	10	25	40
Sliding speed	S	Meter/second	0.5	1.0	1.5

The variable that has to be optimized is the AMCs weight loss. Fifteen experimental trials were designed and are detailed in Table 3.6. An ANOVA analysis was conducted to detect important parameters within a constructed model. The primary objective of ANOVA was to classify the critical parameters and pinpoint the most influential parameter, as well as assess the adequacy of the experimental outcomes. A factor is deemed significant if its corresponding p-value is below 0.05. Minitab software (20.3) was used for regression modelling.

Table 3.6 - Box–Behnken experimental design for dry sliding wear test

Run	Reinforcement (wt.%)	Load (N)	Sliding speed (m/s)
1	10	25	1.0
2	5	40	1.0
3	5	25	0.5
4	15	40	1.0
5	15	10	1.0
6	5	25	1.5
7	10	40	0.5
8	5	10	1.0
9	10	25	1.0
10	10	25	1.0
11	10	40	1.5
12	15	25	0.5
13	10	10	0.5
14	15	25	1.5
15	10	10	1.5

3.13 EROSIVE WEAR TEST

Erosion rate of AMCs were examined using Air Jet Erosion machine. An erosion tester uses high-velocity air streams to expose materials to abrasive particles in order to measure erosion resistance. In this type of test, the specimen is typically placed inside a chamber or enclosure and an abrasive-laden air stream is directed toward the surface of the specimen. In this study, air erosion of aluminium matrix composites is investigated using a Ducom Instruments air jet tester (Model TR 471-600) as shown in Fig. 3.12. This machine can achieve particle velocities ranging from 30 to 100 m/s and the erosion feed rate can range from 1 to 10 g/min. Various angles of impingement are available, including 15°, 30°, 45°, 60°, 75° and 90°. The nozzle diameter is standardized at 1.5 mm. There is an adjustment for the sample size to 20 x 20 x 5 mm. The erodent hopper has a maximum capacity of 2 litres. Temperature options include ambient or up to 600°C.

Fig. 3.12 Air jet erosion tester

The specimen holder utilized for the erosion test procedure is illustrated in Fig. 3.13a and b. An erosion resistance examination was conducted in according to ASTM G76 standards as shown in Fig. 3.13c and d [118]. Before testing, the specimens underwent polishing by emery papers. The study was structured following the BBD methodology to explore the impact of erosive factors on erosion rates. Table 3.7 delineates the test conditions used in solid particle erosion tests.

Experiments were conducted with varying impingement angles, erodent velocity and reinforcement wt.%. To determine the erosion rate of AMCs, the weight of each AMCs sample was recorded both before and after each experiment. This was accomplished using a Wensar electronic balance, which has a precision of ±0.1 mg (Wensar Weighing Scales Limited). The erosion rate of the composites was then calculated using Eq. (3.8).

$$E_r = \frac{(W_1 - W_2)}{T} \quad \textbf{(3.8)}$$

Here, E_r represents the erosion rate in grams per minute, W_1 denotes the specimen's weight before erosion, W_2 indicates the weight after erosion and T signifies the time in minutes.

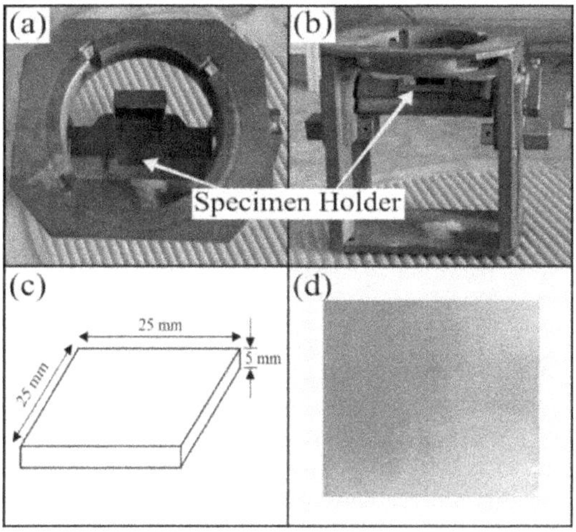

Fig. 3.13 Showing (a) top, (b) front view of specimen holder (c) 2D drawing of erosion sample and (d) erosion specimen

Table 3.7 - Conditions for air jet erosion testing

S.No.	Parameters	Values
1	Erodent material	Alumina (Al_2O_3)
2	Erodent particle size (micron)	45-50
3	Standoff distance (mm)	10
4	Temperature (°C)	Room temperature
5	Erodent feed rate (gm/min)	3
6	Test duration (minute)	5
7	Specimen size (mm)	25x25x5

3.14 REGRESSION MODELLING AND OPTIMIZATION OF EROSIVE WEAR PARAMETERS

The Box-Behnken Design (BBD) was utilized to evaluate the effects of the weight fraction of reinforcement (R), impingement angle (A) and erodent velocity (V) both individually and in combination on the weight loss per minute of the AMCs. Table 3.8 presents the different levels of the parameters used in air jet erosion test. Each parameter was categorized with coding values of -1 for low, 0 for moderate and +1 for high.

Table 3.8 Parameters considered in air jet erosion test process and their levels

Process Parameter	Notation	Unit	Levels		
			-1	0	+1
Reinforcement	R	Wt. %	5	10	15
Impingement angle	A	Degree	30	60	90
Erodent velocity	V	Meter/second	44	72	100

The variable that has to be optimized is the AMCs weight loss. Fifteen experimental trials were designed and are detailed in Table 3.9. An ANOVA analysis was conducted to detect important factors within a constructed model. The primary objective of ANOVA was to classify the critical parameters and pinpoint the most influential parameter, as well as assess the adequacy of the experimental outcomes. A factor is deemed significant if its corresponding p-value is below 0.05. Minitab software (20.3) was used for regression modelling.

Table 3.9 Box–Behnken experimental design for air jet erosion test

Run	Reinforcement (wt.%)	Impingement angle (Degree)	Erodent velocity (m/s)
1	5	30	72
2	10	60	72
3	10	60	72
4	15	90	72
5	10	30	100
6	10	90	100
7	5	60	44
8	10	90	44
9	15	60	44
10	5	90	72
11	10	60	72
12	15	30	72
13	10	30	44
14	15	60	100
15	5	60	100

CHAPTER 4

PHYSICAL, MICROSTRUCTURAL AND MECHANICAL PROPERTIES

4.1 INTRODUCTION

This chapter unveils the profound impact of red mud and sillimanite particles on the physical, microstructural and mechanical properties of the LM25 alloy matrix. The intricacies of density, microstructure, tensile properties, microhardness and thermal conductivity are explored in this chapter. The analysis is divided into three sections: physical properties analysis, microstructural characterization and mechanical properties analysis. To evaluate the performance of the AMCs, their properties were also compared to those of the base alloy.

4.2 PHYSICAL PROPERTIES

4.2.1 Density and Porosity Measurement

Density is a critical material property for many weight-sensitive applications. Consequently, AMCs often replace conventional materials in these applications due to their lower densities. The density of a AMCs relies on the distribution and ratio of the reinforcing and matrix materials. Fig. 4.1 and Table 4.1 depict the theoretical density, experimental density and porosity of the cast composites and the base alloy. The density of composites increases as reinforcement amounts are increased. However, the experimental density was lower than that theoretical density. SR-15 composite has highest density among all the composition and LM25 alloy. The experimental density of the SR-15 composite is 5.68% higher than that of the base alloy. Denser reinforcements embedded in the matrix contribute to the higher density of the composites [119].

The measured density of a composite frequently differs from its theoretical value because of the presence of pores. During the casting process, the reinforcement particles increase the surface area of the slurry exposed to air, leading to the formation of porosity. These pores can significantly impact the mechanical behaviour and overall functioning of the AMCs. Therefore, it is essential to minimize porosity levels. Fig. 4.2 shows a comparison of experimental densities and porosity of base alloy and AMCs. The SR-5 composite exhibits the highest porosity compared to all other composites. Several factors contribute to the increased porosity, including gas entrapment during stirring, hydrogen gas formation, shrinkage during solidification and air bubble infiltration into the slurry [120]. The SR-10 and SR-15 composites have lower porosity than the SR-5 composites. Adding reinforcement particles can make the packing structure denser by filling gaps and voids in the matrix material. As a result, the overall porosity of the AMCs is reduced because the number of empty spaces within the composite is reduced. [121].

Table 4.1 Experimental and theoretical densities of LM25 alloy and AMCs

Material	Density, kg/m³ (X 10³)		
	Theoretical	**Experimental**	**Porosity (vol. %)**
LM-25 Alloy	2.76	2.64±0.081	4.34
SR-5	2.82	2.68±0.056	4.96
SR-10	2.86	2.73±0.039	4.54
SR-15	2.91	2.79±0.092	4.12

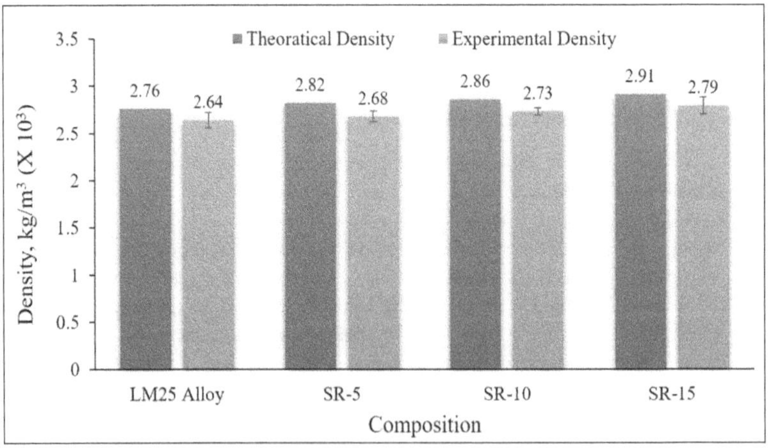

Fig. 4.1 Theoretical and experimental densities graph of LM25 alloy and AMCs

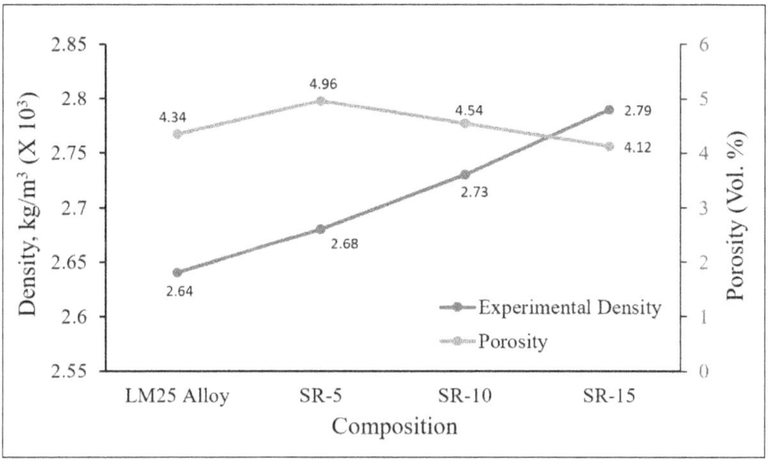

Fig. 4.2 Density and porosity graph as a function of weight percentage of reinforcement

4.2.2 Thermal Conductivity

The thermal conductivity of the AMCs is reported to decreases with increasing amount of reinforcing material as shown in Table 4.2 and Fig. 4.3. The thermal conductivity of the SR-15 composite is decreased by 3.13% compared to the LM25 alloy. Increased relative density of AMCs could lower electron mobility and diminish heat dissipation capability. Sillimanite and red mud exhibit significantly lower thermal conductivity (9.09 W/(m-K) and 0.454 W/(m-K), respectively) in comparison to the LM25 alloy. Therefore, they act as thermal barriers that impede heat flow and reduce the rate of heat transfer [122-124]. The interaction between the matrix and reinforcement particles also influences thermal conductivity. Lower thermal conductivity is also caused by the existence of voids and cavities, which function as thermal barriers at the interface. A low thermal conductivity in processed AMCs indicates strong interfacial bonds [125]. The thermal conductivity of the composites is much higher than the commercially used cast iron (32 W/(m-K) to 45 W/(m-K)) [126]. Automobile components, especially those in the engine and braking systems, generate significant amounts of heat. Materials with good thermal conductivity can quickly dissipate this heat, preventing overheating and ensuring optimal performance. Hence the fabricated composite can be used in manufacturing of various components in automobile industry.

Table 4.2 Thermal conductivity of LM25 alloy and AMCs

Composition	Thermal conductivity (W/m-K)
LM25 Alloy	150.30 ±0.491
SR-5	148.17±0.346
SR-10	147.60±0.400
SR-15	145.59±0.432

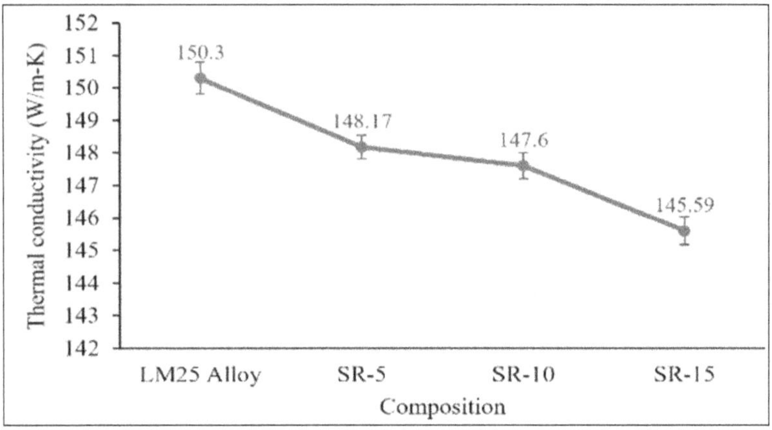

Fig. 4.3 Thermal conductivity graph of LM25 alloy and AMCs

4.3 MICROSTRUCTURAL STUDIES

4.3.1 Optical Microscopy

LM25 is a hypoeutectic alloy composed of a primary silicon phase and an Al-Si eutectic phase. A faceted structure is formed during the solidification of silicon, which is followed by a eutectic phase afterward. Different factors affect the morphology of the primary silicon phase, such as the freezing rate, temperature difference and liquid fraction. Hypoeutectic Al-Si alloy has a cast structure that consists of needle-like eutectic silicon. Primary silicon phases with coarser grain size can lower the alloy's extrudability, machinability, strength and ductility. To reduce the wear rate of a hypoeutectic Al-Si alloy, it is crucial to have an optimal size of primary silicon with consistent distribution [127, 128]. Micrographs of the base alloy are presented in Fig. 4.4a and b at 50X and 200X magnification. The microstructures at various magnifications, specifically at 200X, revealed the arrangement and distribution of silicon precipitates within the eutectic phase, as depicted in Fig. 4.4b. The microstructure of the LM25 alloy depicts that the silicon precipitates in the base alloy are dispersed in the form of needles. Silicon has extremely low solubility in aluminium, causing it to precipitate as nearly pure silicon. In eutectic Al-Si structures, the aluminium and silicon phases undergo competitive growth. [129, 130].

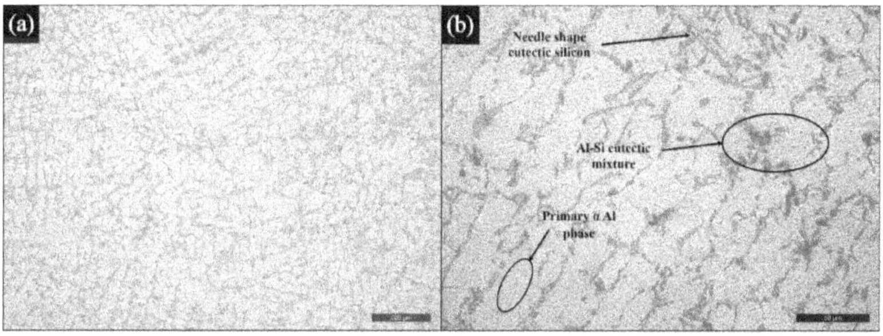

Fig. 4.4 Optical micrographs of LM25 alloy at (a) 50X and (b) 200X magnification

Fig. 4.5a and b, 4.6a and b and 4.7a and b presents the optical micrographs of the SR-5, SR-10 and SR-15 AMCs respectively. Micrographs reveal that there was minimal agglomeration and a fairly uniform distribution of reinforcement particles throughout the matrix. The silicon phase was found to be highly concentrated in the vicinity of the reinforced mineral particles, as illustrated in Fig. 4.6b. This was caused by a thermal mismatch between reinforced particles and LM25 alloy, which restricted the increase in temperature of the reinforcement particles. Consequently, a temperature gradient developed at the boundary between the reinforced particles and the molten LM25 alloy. This temperature difference caused particles to absorb heat from the surrounding matrix and solidify it in their vicinity. The higher melting point of silicon compared to aluminium results in silicon enrichment near the particles during nucleation [131, 132].

Sillimanite and red mud particles exhibit lower thermal conductivity and higher heat diffusivity compared to the LM25 alloy, leading to a reduced cooling rate relative to the molten matrix material. This causes a delay in the solidification process of the matrix. As a result, nucleation of α-Al occurs in regions where the molten AMC is at a lower temperature. Since silicon freezes at a higher temperature than alpha aluminium, it tends to crystallize around the reinforced particles [133, 134].

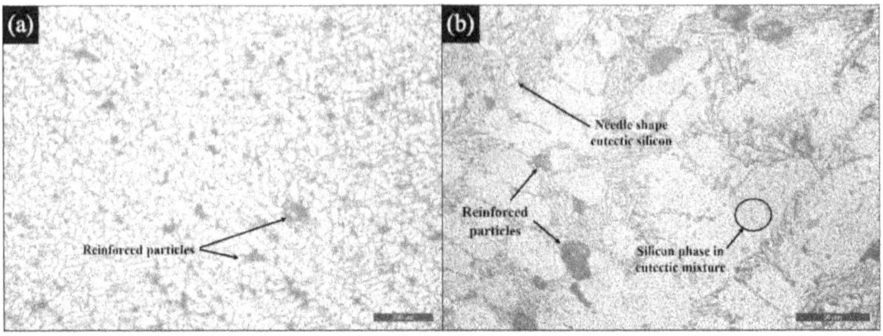

Fig. 4.5 Optical micrographs of SR-5 composite at (a) 50X and (b) 200X magnification

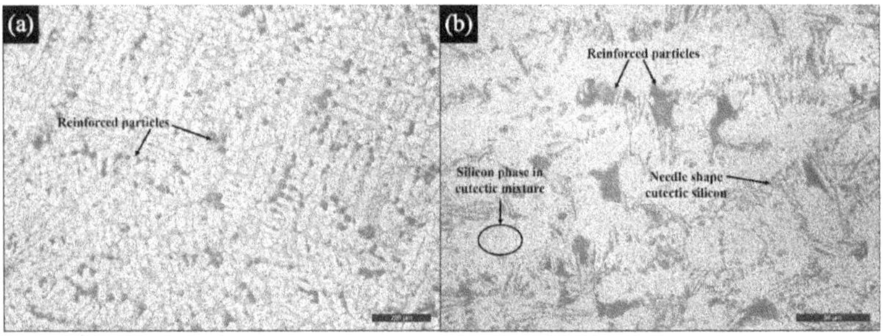

Fig. 4.6 Optical micrographs of SR-10 composite at (a) 50X and (b) 200X magnification

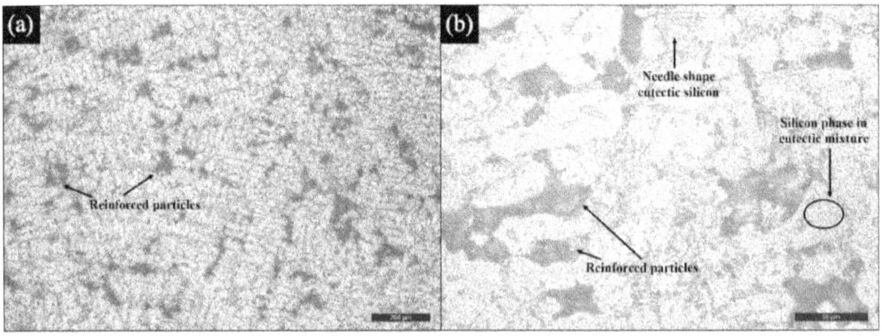

Fig. 4.7 Optical micrographs of SR-15 composite at (a) 50X and (b) 200X magnification

4.3.2 SEM and EDS Analysis

The SEM images of the LM25 alloy and its composites are shown in Fig. 4.8a and Fig. 4.8b–d, respectively. Fig. 4.8a reveals needle-like silicon precipitates within the eutectic phase. The formation of α-Al is primarily attributed to the thermal mismatch between silicon and aluminium, as silicon has significantly lower thermal conductivity compared to aluminium. This thermal disparity causes α-Al to nucleate within the liquid alloy. Fig. 4.8b–d depict a relatively uniform distribution of the reinforced particles. The reinforced particles and the matrix material exhibit a strong interfacial bond, evidenced by the absence of voids at the interface.

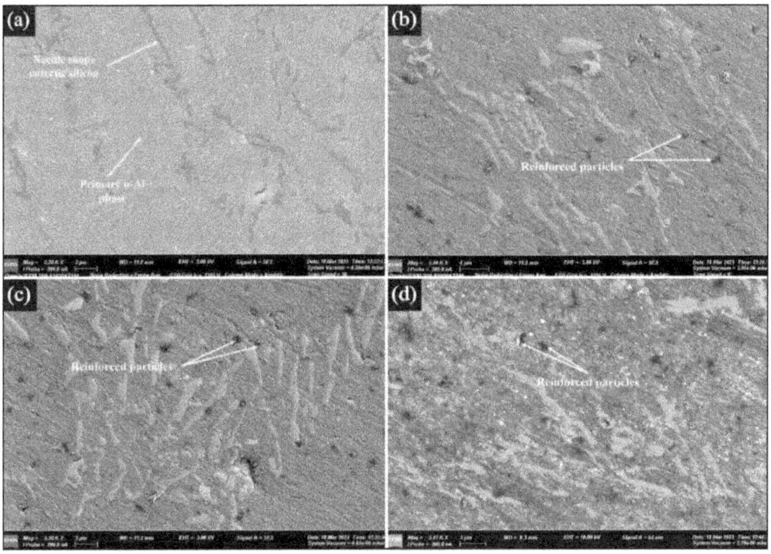

Fig. 4.8 SEM micrographs of (a) LM25 alloy, (b) SR-5 (c) SR-10 and (d) SR-15 composites

Fig. 4.9a-c shows the EDS analysis of the SR-15 composite. The EDS analysis confirms the presence of sillimanite and red mud particles in the base alloy. Fig. 4.9d shows the size of the reinforcing particles.

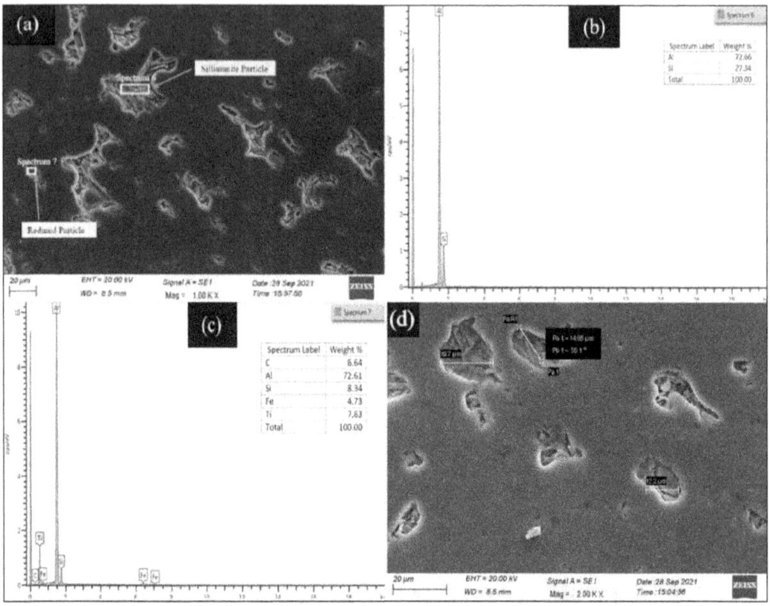

Fig. 4.9 Showing (a) SEM, (b and c) EDS results of SR-15 AMC at 1000X magnification and (d) SEM micrograph presenting reinforcing particle size in SR-15 AMC

4.4 MECHANICAL TESTING

4.4.1 Microhardness Analysis

The hardness of all compositions evaluated at five distinct positions on the surface and an average value is shown in Table 4.3 and Fig. 4.10. The increase in the reinforcement particle amount in the LM25 alloy shows a significant rise in hardness. Since red mud and sillimanite are ceramics with higher hardness values compared to the LM25 alloy, the AMCs become harder as a result. The presence of these reinforced particles creates a barrier to plastic deformation (dislocation movement) caused by the indenter during the hardness test, thus enhancing the hardness of the AMCs [69, 135]. The SR-15 composite shows improvements in hardness of 29.27% compared to LM25 alloy, 16.11% compared to SR-5 composite and 7.14% compared to SR-10 composite. Additionally, the hardness of material impacts its wear rate, as described by Archard's law [136].

Table 4.3 Microhardness of LM25 alloy and AMCs

Composition	VHN
LM25 Alloy	74.12 ± 0.74
SR-5	82.52 ± 2.60
SR-10	89.43 ± 3.80
SR-15	95.82 ± 1.90

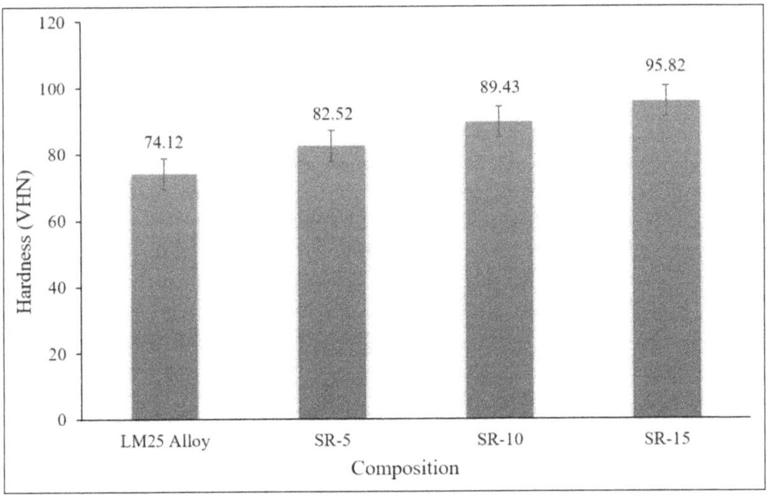

Fig. 4.10 Microhardness graph of LM25 alloy and AMCs

4.4.2 Tensile Test

Table 4.4 and Fig. 4.11 summarize the tensile behaviour of LM25 alloy with different wt.% of reinforcement particles. It is evident that the ultimate tensile strength (UTS), yield strength as well as Youngs modulus increase with an increase in the wt.% of both reinforcements. The SR-15 composite shows a 33.24% increase in UTS, a 59.50% enhancement in yield strength and 10.66 % in the Youngs modulus compared to the LM25 alloy. The enhancement in strength suggests improved bonding of the reinforcement particles with the LM25 alloy. The resistance provided by the reinforcement particles obstructed dislocation movement, thereby increasing tensile strength through Orowan strengthening [137, 138].

Table 4.4. Tensile properties of LM25 alloy and fabricated AMCs

Composition	Ultimate Tensile Strength (N/mm^2)	Yield Strength (N/mm^2)	Youngs Modulus (N/mm^2)	Percentage Elongation	Percentage enhancement in UTS compared to the LM25 alloy
LM25 Alloy	127.022	72	62.787	4.59	-
SR-5	151.163	100.346	65.258	3.40	19
SR-10	161.826	111.446	66.728	2.97	27.39
SR-15	169.256	114.844	69.485	1.20	33.24

Conversely, as reinforcement particle amounts are increased, the elongation or ductility of composites decreases. It is widely recognized that enhancing the yield strength or UTS of AMCs typically reduces their elongation capacity or ductility, particularly when they are reinforced with micron or sub-micron-sized hard particles. The red mud and sillimanite particles enhanced the strength of the AMCs by impeding dislocation movement. This limitation in dislocation movement improved the load transfer capacity between the LM25 alloy and the reinforcement particles [139, 140].

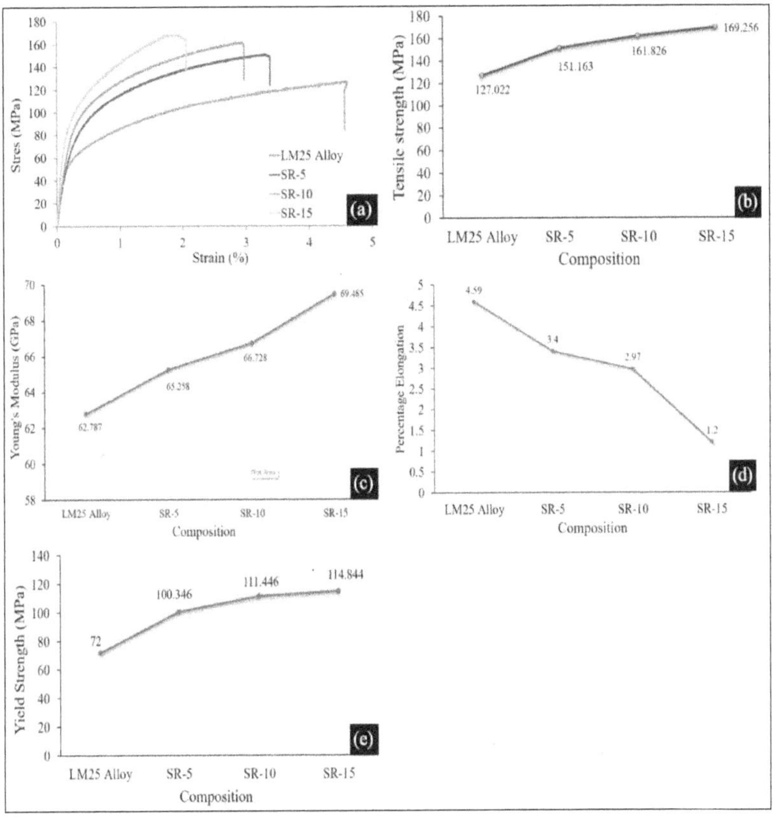

Fig. 4.11 Showing (a) stress-strain behaviour, (b) tensile strength (c) young's modulus, (d) percentage elongation and (e) yield strength of LM25 alloy and aluminium matrix composites

4.4.3 Fractography

Fig. 4.12a-d shows SEM micrographs of fractured surfaces of LM25 alloy, SR-5, SR-10 and SR-15 composite materials after tensile testing. The appearance of equiaxed dimples in LM25 alloy signifies a ductile type of failure, which happens after considerable plastic deformation (Fig. 4.12a). However, the composites exhibited a mixed mode of failure, combining both ductile and brittle fracture, characterized by the presence of cleavage facets and dimples (Fig. 4.12b and c). The small voids depicted in Fig. 4.12d indicate that cracks initiate at the boundary between the LM25 alloy and the reinforced particles through a decohesion process. Enhancing the concentration of reinforcement has a significant effect on the ductility of composites, as evidenced by the reduction in the size of dimples. It is evident that local delamination between the matrix and reinforcement has occurred, manifesting as voids. The SR-15 AMC sample exhibits deep dimples, pointing to failure resulting from the formation, expansion and merging of voids within the matrix [141, 142].

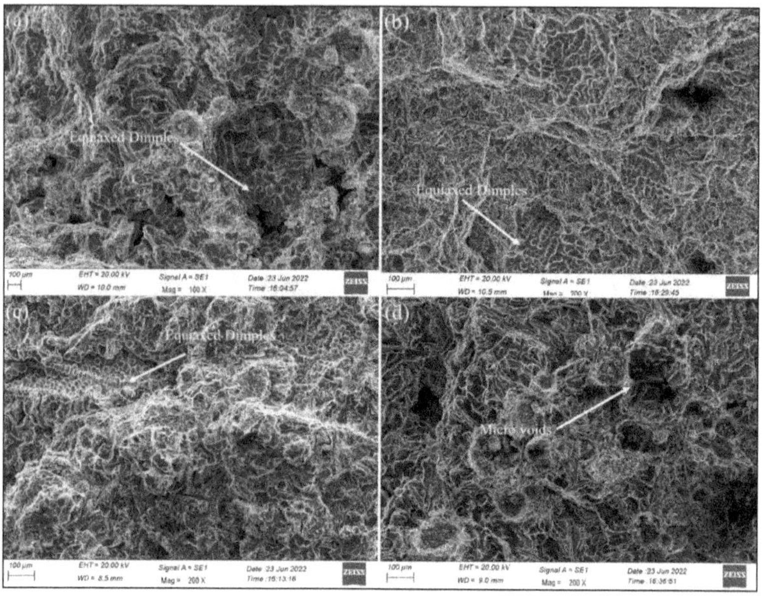

Fig. 4.12 SEM micrographs of broken faces of (a) LM25 alloy, (b) SR-5, (c) SR-10 and (d) SR-15 AMCs

CHAPTER 5

TRIBOLOGICAL PROPERTIES

5.1 INTRODUCTION

This chapter unveils the effect of red mud and sillimanite particles on the dry sliding wear, friction and erosive wear behaviour of the LM25 alloy matrix. The investigation is segmented into two areas: dry sliding wear evaluation with friction and erosive wear analysis. In the first section the impact of sliding speed, applied normal load and wt.% of red mud and sillimanite particles on the dry sliding wear and friction properties of the AMCs at ambient temperature is evaluated through the Box-Behnken Design (BBD) method. In the second section, the effect of erodent velocity, impingement angle and wt.% of red mud and sillimanite particles on erosive wear behaviour of composites at ambient temperature is evaluated through the BBD method. The regression models were formulated for dry sliding wear rate, COF and erosive wear rate. The responses were optimized using the desirability analysis in both sections. A validation test was also conducted with optimized parameters to ensure the accuracy of the optimization process. Further, worn surfaces and eroded surfaces have been studied under SEM for a better understanding of the mechanisms involved in sliding wear and erosive wear.

5.2 DRY SLIDING WEAR AND FRICTION STUDY

Dry sliding wear tests were conducted at ambient temperature on pin on disc machine according to the BBD method. Wear rate and COF measurements were assessed and are detailed in Table 5.1.

Table 5.1 - Box–Behnken experimental design and response for dry sliding wear test

Run	Reinforcement (weight %)	Load (N)	Sliding speed (m/s)	Wear rate (mm^3/m)	COF
1	10	25	1.0	0.0275±0.00282	0.533±0.0216
2	5	40	1.0	0.0334±0.00184	0.624±0.0185
3	5	25	0.5	0.0265±0.00365	0.598±0.0352
4	15	40	1.0	0.0362±0.00295	0.624±0.0185
5	15	10	1.0	0.0221±0.00235	0.362±0.0219
6	5	25	1.5	0.0320±0.00097	0.601±0.0201
7	10	40	0.5	0.0198±0.00153	0.633±0.0119
8	5	10	1.0	0.0266±0.00211	0.476±0.0201
9	10	25	1.0	0.0267±0.00305	0.565±0.0194
10	10	25	1.0	0.0253±0.00295	0.584±0.0185
11	10	40	1.5	0.0284±0.00216	0.585±0.0136
12	15	25	0.5	0.0272±0.00101	0.496±0.0301
13	10	10	0.5	0.0163±0.00116	0.405±0.0155
14	15	25	1.5	0.0245±0.00197	0.503±0.0066
15	10	10	1.5	0.0158±0.00207	0.398±0.0159

5.2.1 Regression Analysis of Wear Rate

The regression analysis was performed using Minitab software (20.3) and the quadratic regression equation was obtained for wear rate as given below in Eq. (5.1).

Wear Rate = 0.02175 - 0.004221 Reinforcement (wt.%) + 0.000251 Load (N) + 0.03714 Sliding Speed (m/s) + 0.000211 Reinforcement (wt.%) *Reinforcement (wt.%) - 0.000010 Load (N)*Load (N) - 0.01690 Sliding Speed (m/s) * Sliding Speed (m/s) + 0.000024 Reinforcement (%) * Load(N) - 0.000820 Reinforcement (wt.%) * Sliding Speed (m/s) + 0.000303 Load (N) * Sliding Speed (m/s) **(5.1)**

The above equation shows the impact of each variable or double interaction on the dry sliding wear rate. A negative coefficient indicates that an individual or double interaction factor negatively impacts the wear rate (i.e., the wear rate decreases), whereas a positive coefficient indicates that the factor increases the wear rate. The equation indicates that reinforcement wt.% had a negative effect on wear rate, whereas sliding speed and load had a positive impact.

Fig. 5.1a and b presents the normal probability graph and predicted vs actual wear rate graph for the experimental data. Fig. 5.1a illustrates the normal probability plot of the residuals, which appear to be normally distributed and align closely with a straight line, showing suitability of the model to the current research. The data were also examined to determine the relationship between the experimental and predicted wear rates, as illustrated in Fig. 5.1b. The predicted values are very close to the experimental values, with all points near the line, indicating the regression model successfully captures the correlation between process variables and response. These findings also indicate that the chosen quadratic model was effective in predicting the response variables for the experimental data.

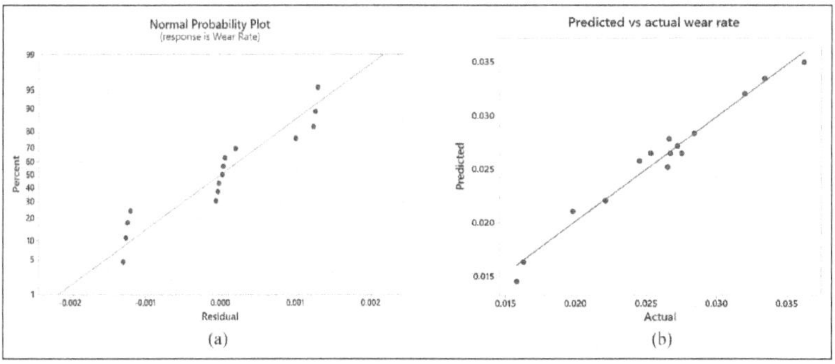

Fig. 5.1 Showing (a) normal probability plot and (b) predicted versus actual wear rate plot

The regression model was analysed through analysis of variance (ANOVA) using the BBD method and Minitab software to determine the adequacy and significance of the model coefficients. The lack of fit tests, sum of squares and overall model statistics results are presented in Table 5.2. The model statistics shows, the p-value for the quadratic model is 0.002 which is much below 0.05. A p-value below 0.05 indicate that the factors being tested have a notable impact on the outcomes, indicating that the model was statistically considerable at the 95% confidence level [143]. The R^2 value, also called the coefficient of determination, represents the closeness of data points to the regression line and also indicate the adequacy of the model. The R^2 value close to 100% indicates a better fit of the experimental data, while a lower R^2 value suggests a less accurate fit [144]. Table 5.2 indicates that R^2 and adjusted R^2 value responses (wear rate) are both 97.36% and 92.62%, which indicate a very good relationship between the input variables and the responses. The F-value for the regression model is significantly high at 20.51. A higher F value suggests that the differences between group means are more substantial compared to the differences within the groups (residual variation), suggesting that the model provides a better fit to the data [145]. Thus, the quadratic model effectively describes the relationship between the independent variables and the response. Applied normal load has the major impact on the wear rate, followed by sliding speed and reinforcement wt.%.

Table 5.2 - ANOVA of dry sliding wear rate

Source	DF	Adjusted SS	Adjusted MS	F value	P value
Model	9	0.000448	0.00005	20.51	0.002
Linear	3	0.000195	0.000065	26.81	0.002
Reinforcement (wt.%)	1	0.000009	0.000009	3.72	0.111
Load (N)	1	0.000171	0.000171	70.58	0
Sliding Speed (m\s)	1	0.000015	0.000015	6.13	0.056
Square	3	0.000202	0.000067	27.73	0.002
Reinforcement (wt.%) *Reinforcement (wt.%)	1	0.000103	0.000103	42.38	0.001
Load (N)*Load (N)	1	0.000018	0.000018	7.37	0.042
Sliding Speed (m\s) * Sliding Speed (m\s)	1	0.000066	0.000066	27.18	0.003
2-Way Interaction	3	0.000051	0.000017	6.99	0.031
Reinforcement (wt.%) *Load (N)	1	0.000013	0.000013	5.49	0.066
Reinforcement (wt.%) *Sliding Speed (m\s)	1	0.000017	0.000017	6.93	0.046
Load (N) * Sliding Speed (m\s)	1	0.000021	0.000021	8.54	0.033
Error	5	0.000012	0.000002		
Lack of Fit	3	0.00001	0.000003	2.59	0.291
Pure Error	2	0.000002	0.000001		
Total	14				

Summary of Model

S	R^2	R^2 (adjusted)	R^2 (predicted)
0.0015571	97.36%	92.62%	65.22%

Fig. 5.2 illustrates the main effect plots for mean wear rate showing the variation against all wear parameters. The graph illustrates a parabolic correlation between sliding speed and reinforcement, with the mean of wear rate. The wear rate of AMC decreases as the reinforcement amount increases because the material becomes harder. Ceramic

materials are much harder than the base alloy and when these hard reinforcements are added to the matrix, they make the overall composite harder. Harder materials are more resistant to plastic deformation and abrasion, leading to a lower wear rate. Initially the wear rate decreases up to 10 wt.% reinforcement due to increase in hardness thereafter wear rate increases. However, SR-15 composite has lower wear rate than 5 wt.% composite. When the particles are deformed, they separate from the aluminium matrix through plastic deformation. Increasing wear rates result from these particles ploughing away from the composite face. The particles between the disc and pin contribute to increased wear. With a higher reinforcement percentage, these particles are more prone to breaking and being displaced [146].

The mean of wear rate tends to increase with higher sliding speeds, as the frictional heat generated by the sliding motion causes the metal matrix to soften. However, it begins to reduce after reaching 1 m/s which may be due to rapid formation of tribolayers at high sliding speed. The tribolayers at high sliding speeds can reduce wear rate by altering the surface properties and enhancing the durability of the sliding interface [147].

In Fig. 5.2, it is shown that an increase in load results in a higher wear rate. The application of high loads results in greater wear rates as disturbed tribolayers and enlarged contact zones contribute to accelerated wear. Plastic deformation causes layers within the specimen to deform, leading to the growth of subsurface cracks and subsequent surface delamination [86].

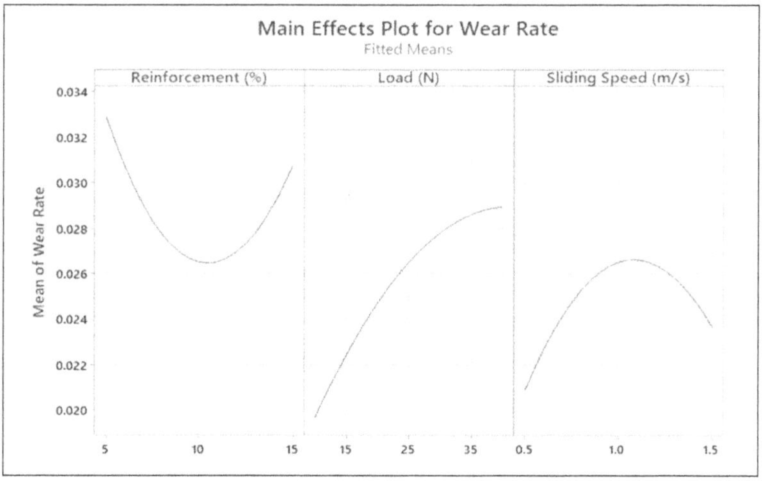

Fig. 5.2 Main effects plot for wear rate as a function of reinforcement (wt. %), load (N) and sliding speed (m/s)

MML's stability might be maintained under low loading conditions, but as the load rises, it could become unstable and show more wear as shown in Fig. 5.3 [147].

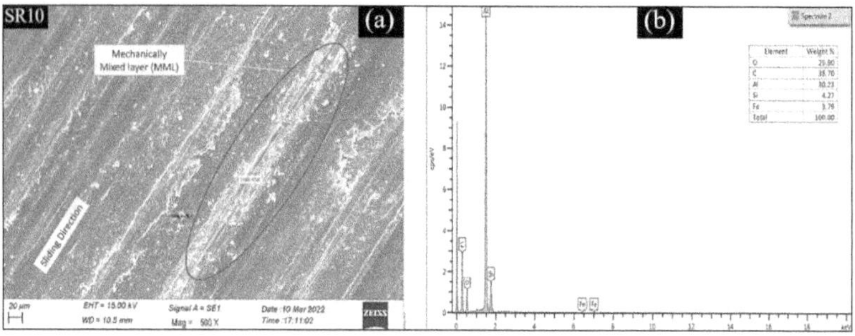

Fig. 5.3 Showing (a) SEM and (b) EDS image of wear track of SR-10 composite at 40N load and 1.5 m/s velocity

The surface plot in Fig. 5.4 represents the wear rate corresponding to each pair of process parameters. The wear rate versus reinforcement and sliding speed plot, holding the load constant at 25 N, reveals that the wear rate drops as the reinforcement wt.% increases. Additionally, at lower sliding speeds, the wear rate is higher and decreases as sliding speed increases to a certain point, indicating a non-linear relationship. The minimum wear rate appears at moderate reinforcement and sliding speed values, suggesting an optimal combination of these factors for reducing wear.

The wear rate versus load and sliding speed plot, holding reinforcement constant at 10 wt.%, indicates that the wear rate increases with an increase in load. The wear rate is higher at lower sliding speeds and decreases as the sliding speed increases to a point. The significant impact of load on the wear rate is evident, especially at higher values, with a saddle point or minimum wear rate at a specific combination of load and sliding speed.

The wear rate versus reinforcement and load plot, while holding the sliding speed fixed at 1 m/s shows that the wear rate falls with an enhancement in reinforcement wt.% and increases with an increase in load. This plot suggests that high reinforcement and low load lead to the lowest wear rates, highlighting a clear interaction between reinforcement and load.

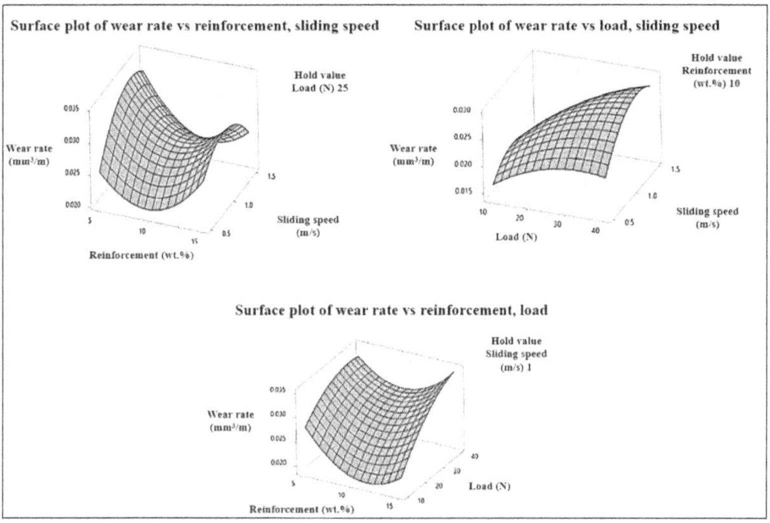

Fig. 5.4. Surface plots of wear rate versus reinforcement (wt.%), load (N) and sliding speed (m/s)

5.2.2 Regression Analysis of COF

The regression analysis was performed using Minitab software (20.3) and the quadratic regression equation was obtained for COF as given below in Eq. (5.2).

COF= 0.383 - 0.0198 Reinforcement (wt.%) + 0.01371 Load (N) + 0.129 Sliding Speed (m/s)+ 0.000102 Reinforcement (wt.%)*Reinforcement (wt.%)- 0.000185 Load (N)*Load (N)-0.0548 Sliding Speed (m/s)*Sliding Speed (m/s)+ 0.000380 Reinforcement (wt.%)*Load (N) + 0.00040 Reinforcement (wt.%)*Sliding Speed (m/s) -0.00137 Load (N) * Sliding Speed (m/s) **(5.2)**

The equations above demonstrate the effect of each variable or interaction on the COF. A negative coefficient indicates that an individual variable or interaction decreases the COF, while a positive coefficient signifies an increase in the COF. Specifically, the equation shows that reinforcement wt.% has a negative effect on the COF, whereas sliding speed and load have a positive impact.

The normal probability plot and predicted vs actual COF graph, based on the experimental data, are presented in Fig. 5.5a and b. In Fig. 5.5a, the normal probability plot of residuals shows that points are normally distributed and closely align with a straight line, indicating the model's suitability for the current research. Fig. 5.5b examines the relationship between experimental and predicted COF, revealing that the predicted values closely match the experimental values, with all points clustering near the line. This demonstrates that the regression model effectively captures the correlation between process variables and response. These results affirm the efficacy of the chosen quadratic model in predicting the response variables for the experimental data.

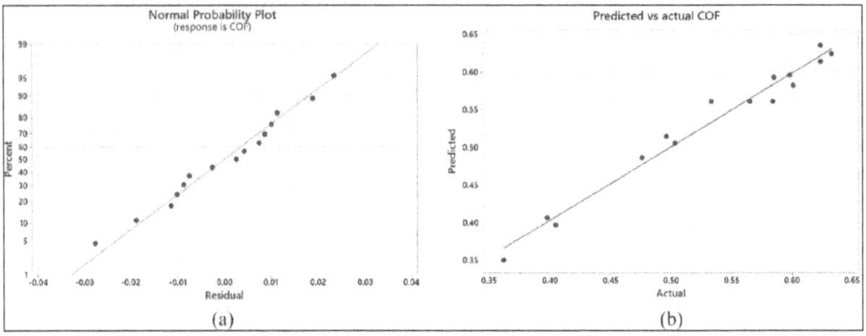

Fig. 5.5 Showing (a) normal probability plot and (b) predicted versus actual plot of coefficient of friction

The regression model was assessed utilizing ANOVA based on the BBD method with Minitab software to evaluate the fitness and significance of the model coefficients. The lack of fit tests, sum of squares and overall model statistics results are detailed in Table 5.3. According to the model statistics, the p-value for the quadratic model is 0.002, which is less than 0.05. The regression model is shown to be statistically significant at the confidence level of 95% according to this value. Table 5.3 indicates that R^2 and adjusted R^2 value responses (COF) are both 97.49% and 92.98%, which highlights a significant relationship between the input parameters and the response. Furthermore, the regression model exhibits a notable F-value of 21.61. Consequently, the quadratic model offers a thorough explanation of the correlation between the independent variables and their resulting responses. Applied normal load has the most significant impact on the COF, followed by reinforcement weight percentage and sliding speed.

Table 5.3 - ANOVA of coefficient of friction

Source	DF	Adjoint SS	Adjoint MS	F Value	P Value
Model	9	0.108319	0.012035	21.61	0.002
Linear	3	0.097656	0.032552	58.45	0.000
Reinforcement (wt.%)	1	0.012324	0.012324	22.13	0.005
Load (N)	1	0.085078	0.085078	152.78	0.000
Sliding Speed (m/s)	1	0.000253	0.000253	0.45	0.530
Square	3	0.006990	0.002330	4.18	0.079
Reinforcement (%)*Reinforcement (wt.%)	1	0.000024	0.000024	0.04	0.844
Load (N)*Load (N)	1	0.006423	0.006423	11.53	0.019
Sliding Speed (m/s)*Sliding Speed (m/s)	1	0.000694	0.000694	1.25	0.315
2-Way Interaction	3	0.003673	0.001224	2.20	0.206
Reinforcement (wt.%)*Load (N)	1	0.003249	0.003249	5.83	0.060
Reinforcement (wt.%)*Sliding Speed (m/s)	1	0.000004	0.000004	0.01	0.936
Load (N) * Sliding Speed (m/s)	1	0.000420	0.000420	0.75	0.425
Error	5	0.002784	0.000557		
Lack-of-Fit	3	0.001456	0.000485	0.73	0.622
Pure Error	2	0.001329	0.000664		

Summary of Model

S	R^2	R^2 (adjoint)	R^2 (predicted)
0.0235984	97.49%	92.98%	76.35

The effect of load, sliding speed and reinforcement wt.% on the mean of COF of composites is depicted in Fig. 5.6. According to the plot, the COF value ascends with an increase in load. Increasing the normal load leads to a greater frictional force between the specimen and the rotating disc material, validating the friction law. Fig. 5.6 illustrates that the COF steadily declines as the weight percentage of reinforcement particles increases. The hard ceramic dispersoid acts as a protrusion on the contact surface of the pin. These protrusions absorb the most pressure and defend against the abrasive particles penetrating the pin surface. Additionally, the elevated hardness of the composite plays a role in the decreased COF value [146]. Fig. 5.6 presents the variation in the COF of specimen with respect to sliding speed. The friction coefficient goes up with increasing sliding speed, up to a maximum of 0.8 m/s. This is due to more abrasive wear caused at initial dry sliding conditions and then decreases as the sliding speed continues to increase. The COF might decrease if the material softens at an accelerated rate due to rising temperatures between the specimen and the disc [148].

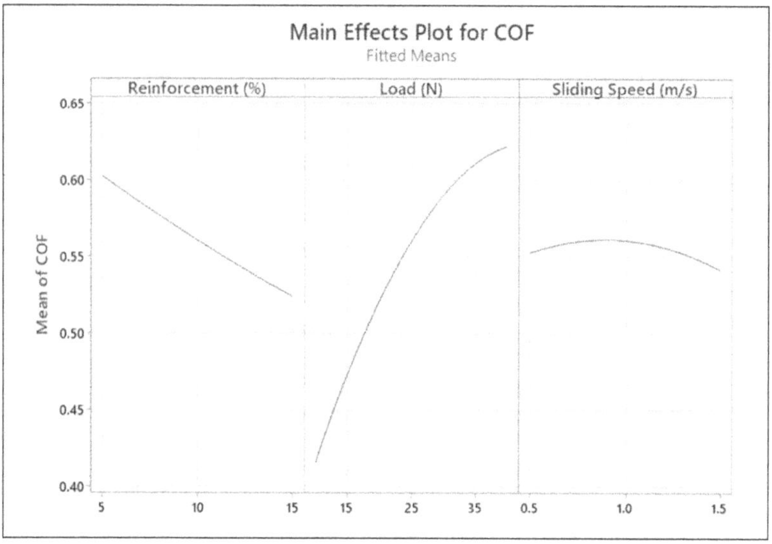

Fig. 5.6 Main effects plot for COF as a function of reinforcement (wt.%), load (N) and sliding speed (m/s)

The surface plots in Fig. 5.7 depicts the COF in relation to each combination of process parameters. The COF versus reinforcement and sliding speed plot, holding the load constant at 25 N, shows that as the reinforcement wt.% and sliding speed increase, the COF decreases. The combined effect shows a more significant reduction in COF at higher reinforcement levels and sliding speeds. This indicates that both increasing reinforcement and sliding speed can effectively reduce the COF under a constant load of 25 N.

The COF versus reinforcement wt.% and load plot, while holding the sliding speed fixed at 1 m/s, shows that as the reinforcement weight percentage increases, the COF decreases, while as the load increases, the COF increases. The combined effect shows that the decrease in COF from added reinforcement is partially offset by the increase in COF under higher loads. However, at higher reinforcement levels, the COF remains relatively low despite increasing loads. This suggests that reinforcement is more effective in controlling COF even under varying loads when the sliding speed is constant at 1 m/s.

The COF vs load and sliding speed plot, holding reinforcement constant at 10 wt.%, indicates that as the sliding speed increases, the COF decreases, while as the load increases, the COF increases. The combined effect shows that the COF is minimized at higher sliding speeds and lower loads. With a constant reinforcement of 10 wt.%, increasing the sliding speed significantly reduces the COF, while increasing the load increases the COF. This indicates that higher sliding speeds are effective in reducing friction even when the load increases.

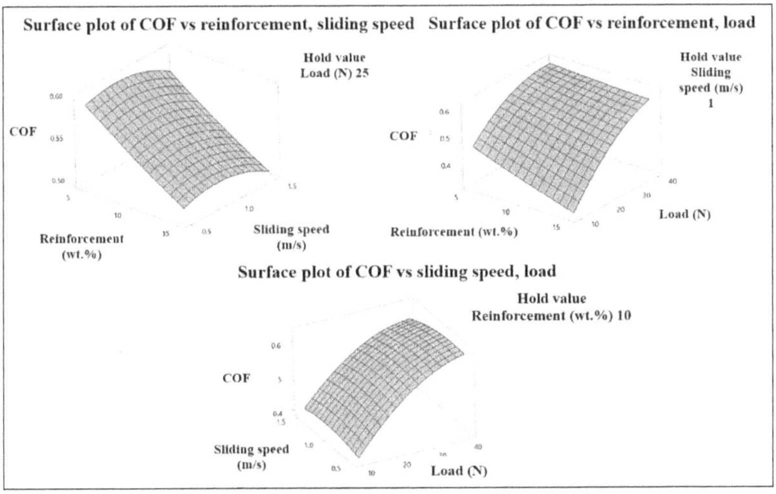

Fig. 5.7 Surface plots of COF versus reinforcement (wt.%), load (N) and sliding speed (m/s)

5.2.3 Optimization of Wear Rate and COF

The optimal input parameters were determined by optimizing the responses to reduce the wear rate and COF of the developed AMCs. This optimization aims to reduce wear rate and COF by optimizing parameters, including load, sliding speed and reinforcement wt.%. Fig. 5.8 depicts a desirability analysis conducted in Minitab, showing the optimization of three factors (reinforcement wt.%, load and sliding speed). The overall high composite desirability (1.000) indicates that the selected values for reinforcement wt.%, load and sliding are optimal for minimizing both the wear rate and the COF in this study. The optimal reinforcement level of 13.48 wt.%, with a load of 10

N and a sliding speed of 1.5 m/s, resulted in a wear rate of 0.01366 mm^3/m and a friction coefficient of 0.3615.

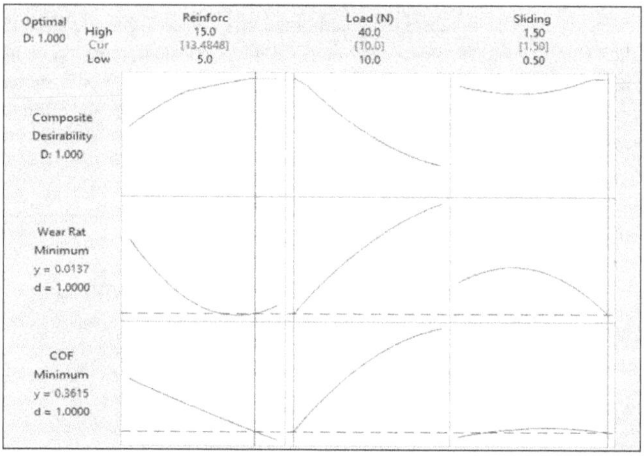

Fig. 5.8 Optimal response values for reinforcement (wt.%), load (N) and sliding speed (m/s)

5.2.4 Confirmation Test

A validation test was conducted to ensure the precision of the optimization procedure. The composite was developed through a like process, containing 13.48 wt.% reinforcement in a 1:1 ratio. A wear test was conducted on the optimum conditions to determine the wear rate and COF. Experimental values of wear rate and COF were compared with the model predicted values as shown in Table 5.4. The error being under 6% is a significant factor, making the proposed method suitable for wear analysis forecasting.

Table 5.4- Results of the confirmation test for dry sliding wear test

Reinforcement (wt.%)	Load (N)	Sliding speed (m/s)	Experimental values		Modelled values		Error (%)	
			Wear Rate (mm^3/m)	COF	Wear Rate (mm^3/m)	COF	Wear Rate (mm^3/m)	COF
13.48	10	1.5	0.0145	0.354	0.01366	0.3615	5.79	2.11

5.2.5 Worn Surface Topography

The wear track of SR-15 at loads of 10 N, 25 N and 40 N, with a sliding speed of 1.5 m/s, is depicted in Fig. 5.9a-c. Abrasive grooves can be seen along the sliding direction at low loads, which indicates an abrasive wear mechanism (Fig. 5.9a). The application of a 25 N load results in various abrasive grooves parallel to the direction of sliding and delamination is observed at particular sites, demonstrating both adhesive and delamination wear phenomena (Fig. 5.9b). When the load and sliding speed are elevated, frictional forces cause the interface temperature to rise. This temperature spike makes the material softer, which results in localized deformation. Moving from adhesion to severe delamination raises the wear rate. Higher contact loads cause the pin grooves to become more pronounced (Fig. 5.9c). The presence of hard particles trapped between the pin and disc results in the creation of abrasive grooves [149]. The sharpness of abrasives enables them to remove material via micro-ploughing. The resulting abrasive grooves are formed from the repeated sliding of these micro-ploughing traces [150].

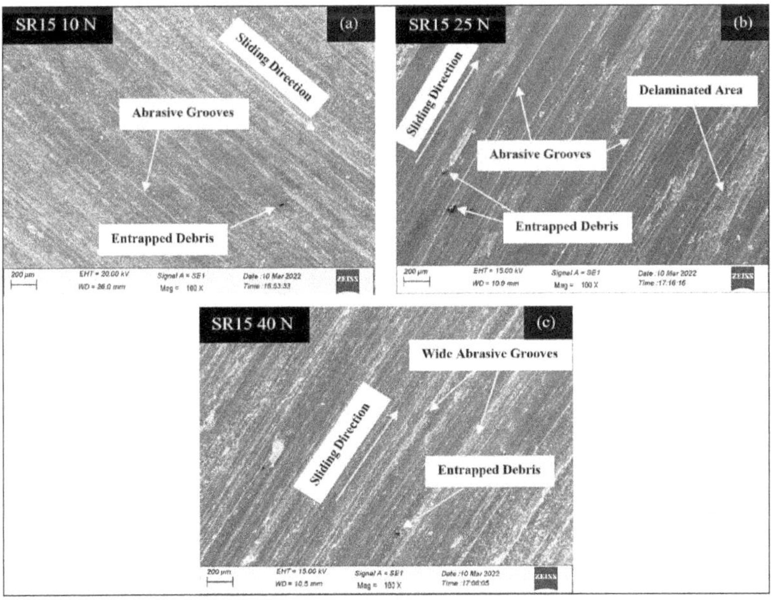

Fig. 5.9 Worn surface topography of SR-15 sample at (a) 10 N, (b) 25 N and (c) 40 N load and 1.5 m/s sliding speed

Fig. 5.10a-c represents the wear track of SR-5, SR-10 and SR-15 composites at 40N load and 1.5 m/s speed. The wear pattern on the SR-15 composite reveals thin grooves that widen with decreasing reinforcement. This indicates that as the percentage of reinforcement particles increases, the composites exhibit greater abrasion resistance and improved load-bearing capacity.

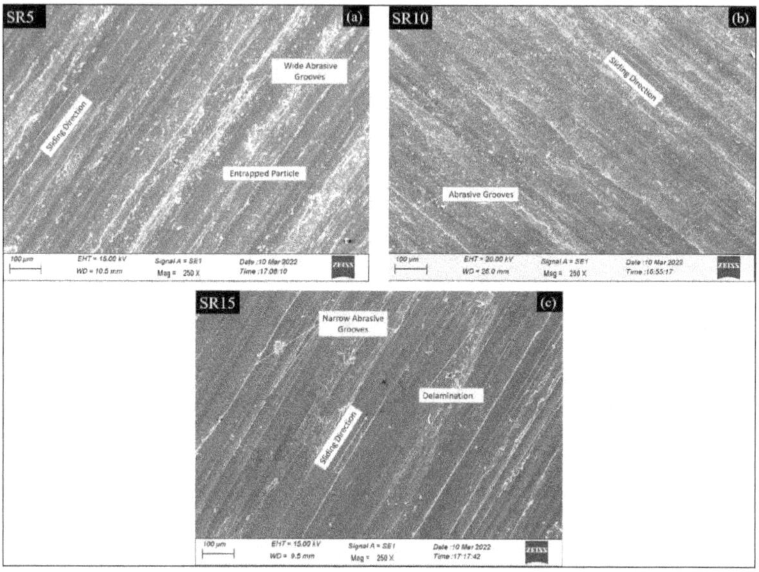

Fig. 5.10. Worn surface topography of (a) SR-5, (b) SR-10 and (c) SR-15 specimens at 40 N load and 1.5 m/s velocity

5.3 EROSIVE WEAR ANALYSIS

Erosion tests were conducted at room temperature following the Box-Behnken Design matrix. Erosion rate values were calculated and presented in Table 5.5.

Table 5.5 Box–Behnken experimental design and response for air jet erosion test

Run	Impingement angle (deg.)	Erodent velocity (m/s)	Reinforcement (wt.%)	Erosion rate (gm/min)
1	90	100	10	0.00866
2	60	72	10	0.00338
3	60	72	10	0.00442
4	60	100	5	0.00896
5	60	44	15	0.00370
6	60	100	15	0.00718
7	90	72	15	0.00667
8	60	44	5	0.00218
9	30	72	5	0.00418
10	30	100	10	0.00698
11	30	72	15	0.00492
12	90	44	10	0.00201
13	60	72	10	0.00442
14	90	72	5	0.00438
15	30	44	10	0.00238

5.3.1 Regression Analysis of Erosion Rate

The regression analysis was performed using Minitab software (20.3) and the quadratic regression equation was obtained for erosion rate as given below in Eq. (5.3).

Erosion rate (gm/min)=0.00408 - 0.000087 Impingement angle (deg.) - 0.000010 Erodent velocity (m/s) - 0.000246 Reinforcement (wt.%) + 0.000000 Impingement angle (deg.) *Impingement angle (deg.) + 0.000001 Erodent velocity (m/s)*Erodent velocity (m/s) + 0.000029 Reinforcement (wt.%)*Reinforcement (wt.%) + 0.000001 Impingement angle (deg.)*Erodent velocity (m/s) + 0.000003 Impingement angle (deg.)*Reinforcement (wt.%) - 0.000006 Erodent velocity (m/s)*Reinforcement (wt.%)

(5.3)

The normal probability plot and the graph of predicted versus actual erosion rates, derived from the experimental data, are illustrated in Fig. 5.11a and b respectively. In Fig. 5.11a, the normal probability plot of residuals shows that points are normally distributed and closely align with a straight line, indicating the model's suitability for the current research. Fig. 5.11b examines the relationship between experimental and predicted erosion rate, revealing that the predicted values closely match the experimental values, with all points clustering near the line. This demonstrates that the regression model effectively captures the correlation between process variables and response. These results affirm the efficacy of the chosen quadratic model in predicting the response variables for the experimental data.

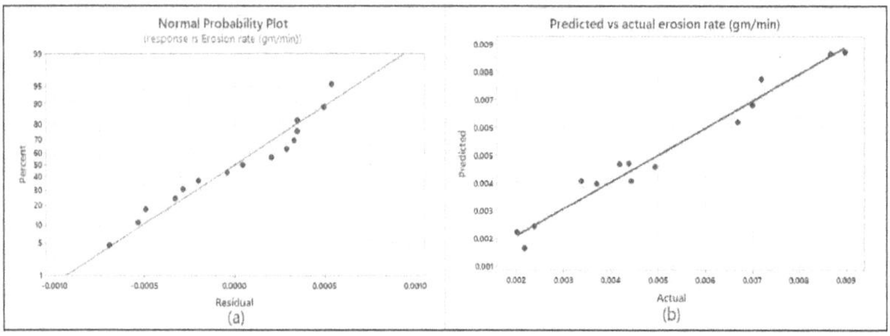

Fig. 5.11 Showing (a) normal probability plot and (b) predicted versus actual plot of erosion rate

The regression model was assessed using analysis of variance (ANOVA) based on the BBD with Minitab software to evaluate the fitness and significance of the model coefficients. The lack of fit tests, sum of squares and overall model statistics results are presented in Tables 5.6. According to the model statistics, the p-value for the quadratic model is 0.003 which is less than 0.05. A lower p-value, indicating that the model is significant at the 95% confidence level. Table 5.6 indicates that R^2 and adjusted R^2 value responses (erosion rate) are both 96.80% and 91.04%, which indicates a strong association between the input data and the responses generated.

Furthermore, the regression model exhibits a notable F-value of 16.81. A higher F-value suggests that the variation among group means exceeds the variation within the groups (residual variation), underscoring the model's suitability for the data. Consequently, the quadratic model offers a thorough explanation of the correlation among the input parameters and the associated outcome. Erodent velocity is the most significant factor among all the input parameters employed for the erosive wear test, followed by the impingement angle and reinforcement wt.%.

Table 5.6 - ANOVA of erosion rate

Source	DF	Adjusted SS	Adjusted MS	F Value	P Value
Model	9	0.000068	0.000008	16.81	0.003
Linear	3	0.000060	0.000020	44.55	0.000
Impingement angle (deg.)	1	0.000001	0.000001	2.96	0.146
Erodent velocity (m/s)	1	0.000058	0.000058	128.55	0.000
Reinforcement (wt.%)	1	0.000001	0.000001	2.14	0.204
Square	3	0.000004	0.000001	2.65	0.160
Impingement angle (deg.) *Impingement angle (deg.)	1	0.000000	0.000000	0.45	0.532
Erodent velocity (m/s) *Erodent velocity (m/s)	1	0.000002	0.000002	4.03	0.101
Reinforcement (wt.%) *Reinforcement (wt.%)	1	0.000002	0.000002	4.38	0.090
2-Way Interaction	3	0.000004	0.000001	3.24	0.119
Impingement angle (deg.) *Erodent velocity (m/s)	1	0.000001	0.000001	2.33	0.188
Impingement angle (deg.) *Reinforcement (wt.%)	1	0.000001	0.000001	1.34	0.300
Erodent velocity (m/s) *Reinforcement (wt.%)	1	0.000003	0.000003	6.05	0.057
Error	5	0.000002	0.000000		
Lack-of-Fit	3	0.000002	0.000001	1.41	0.440
Pure Error	2	0.000001	0.000000		
Total	14	0.000070			

Summary of Model

S	R^2	R^2 (adjusted)	R^2 (predicted)
0.0006706	96.80%	91.04%	62.93%

The relationship between the mean erosion rate and all input factors is depicted in Fig. 5.12. Plots shows a parabolic relationship between all three parameters (impingement angle, erodent velocity and reinforcement wt.%) and with mean of erosion rate. Erodent velocity is most critical parameter influencing the erosion rate of composite. It is widely recognized that increasing velocity enhances the kinetic energy of erodent particles, leading to increased material removal from the target surface. An increase in velocity not only enhances kinetic energy but also extends the time between particle impacts on the surface. This behavior results in a greater extent of material extraction [151]. The primary reasons for increased material removal at higher velocities include the greater stress experienced by the target surface, even at low angles of contact. The surface undergoes wear due to the abrasive action of erodent particles at low contact angles. An increasing angle of impingement leads to material removal through plastic deformation. At elevated impingement angles, brittle fractures can occur in the material, leading to an elevated erosion rate [152]. The relationship of erosion rate to the velocity of erodent is illustrated by a power law in the Eq. (5.4) [153].

$$E = kV^n \qquad \textbf{(5.4)}$$

Here, E represents erosion rate at steady-state, V denotes the particle velocity, k is a constant and n is the velocity exponent. Generally, materials exhibiting ductile behavior typically have a value of n between 2 and 3, while materials that display brittle characteristics usually have a value of n between 3 and 5.

The exponent value indicates that the material removal process in pure aluminium primarily results from plastic deformation. Compared to its composites, aluminium is inherently more ductile. When erodent particles strike the aluminium surface, they induce work hardening. This work hardening effect creates an anvil-like resistance in the material, which helps to reduce the erosion rate at lower velocity of erodent material [69].

Increasing the impingement angle leads to a higher erosion rate in AMCs. With higher impingement angles, material removal increases due to mechanisms such as cutting, indentation and crater formation. In contrast, at low angles, material removal is primarily due to the deflection of the target surface. For ductile materials, the highest level of erosion is observed at an impingement angle between 20 and 30 degrees. Conversely, for brittle materials, the maximum erosion occurs at a 90° angle [154].

The erosion rate decreases as the reinforcement amount increases but up to certain limit after it starts increases. The reinforcement materials act as barriers that resist the penetration and cutting action of erodent particles which results in lower erosion rate. The SR-15 composite has the maximum erosion rate. As the reinforcement is increases the material becomes brittle and this causes fractures and chipping due to the material's low capacity to deform plastically [155].

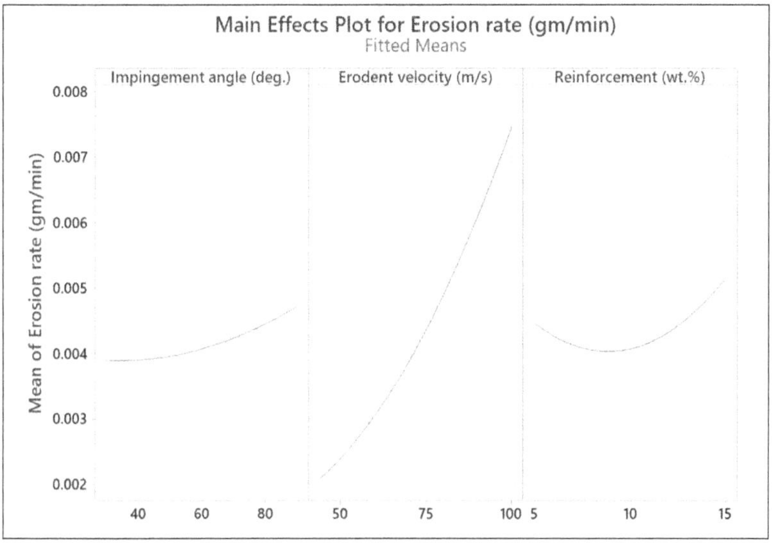

Fig. 5.12 Main effects plot for erosion rate (gm/min) as a function of impingement angle (deg.), erodent velocity (m/s) and reinforcement (wt.%)

The erosion rate for each combination of process parameters were plotted, as depicted in Fig. 5.13. The erosion rate versus erodent velocity and impingement angle surface plot, holding reinforcement at 10 wt.%, shows that the highest erosion rate occurs at high impingement angles and high erodent velocities. The erosion rate tends to decrease as either the impingement angle increases or the erodent velocity decreases.

The erosion rate versus reinforcement and impingement angle surface plot, holding erodent velocity constant at 72 m/s, reveals that an increase in reinforcement percentage and impingement angle both contribute to a reduced erosion rate, with the lowest rates occurring at high reinforcement and large angles.

The erosion rate versus reinforcement and erodent velocity surface plot, holding impingement angle constant at 60° highlights that both higher reinforcement percentages and higher erodent velocities result in lower erosion rates.

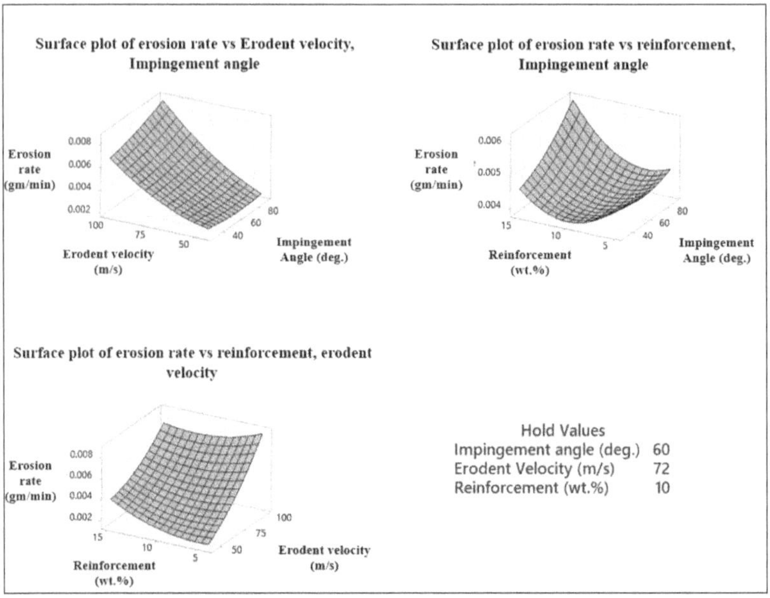

Fig. 5.13 Surface plots of erosion rate versus impingement angle (deg.), erodent velocity (m/s) and reinforcement (wt.%)

5.3.2 Optimization of Erosion Rate

The ideal process parameters were determined through response optimization (desirability analysis) to reduce the erosion rate of the AMCs. Fig. 5.14 shows a desirability analysis conducted in Minitab, revealing the optimization of three parameters (reinforcement wt.%, erodent velocity and impingement angle). The composite desirability score of 1.000 indicates that the selected values for these factors effectively minimize erosion rate in this study. Reinforcement of 5 wt.%, erodent velocity of 1.5 m/s and impingement angle of 90° were found optimal for achieving an erosion rate of 0.0013883 gm/min.

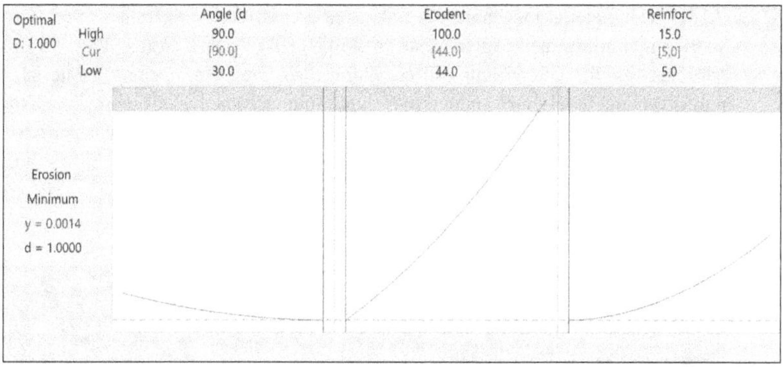

Fig. 5.14 Optimum response value of impingement angle (deg.), erodent velocity (m/s) and reinforcement (wt.%)

5.3.3 Confirmation Test

A validation test was carried out to verify the accuracy of the optimization method. The composite undergoes testing using optimal parameters. Table 5.7 presents the experimental results based on these optimal parameters, comparing them with the optimized responses of the model. Since the error rate is under 6%, the proposed technique is reliable enough to serve as a predictive tool for wear analysis.

Table 5.7- Results of the confirmation test for erosion test

Reinforcement (wt. %)	Erodent Velocity (m/s)	Impingement angle (deg.)	Experimental values of Erosion rate (gm/min)	Modelled values of Erosion rate (gm/min)	Error (%)
5	44	90	0.00147	0.0013883	5.55

5.3.4 Eroded Surface Analysis

Fig. 5.15a-c illustrates the eroded samples of SR-5, SR-10 and SR-15 at a velocity of 72 m/s and 60° impingement angle. The SR-15 composite has the higher erosion rate due to the reduced space between reinforcement particles. The brittle fracture in the reinforcement propagates, which in turn causes microcracking in adjacent reinforcements. This process leads to the loosening and detachment of particles from the matrix material. It is also observed that in the SR-15 reinforced composites, a large quantity of reinforcement particles was taken out from the workpiece material (Fig.

5.15c). The material mechanism is consistent across AMCs regardless of reinforcement wt.% and impingement angle. The erosion mainly occurs around the reinforcement particles, as the aluminium is less hard than the reinforcement material. When reinforcement levels are higher, the aluminium particles erode first, weakening the binding energy of the reinforcement. This results in the reinforcement particles being extracted instead of breaking into pieces. This further contributes to the increased erosion rate as observed in the SR-15 composite.

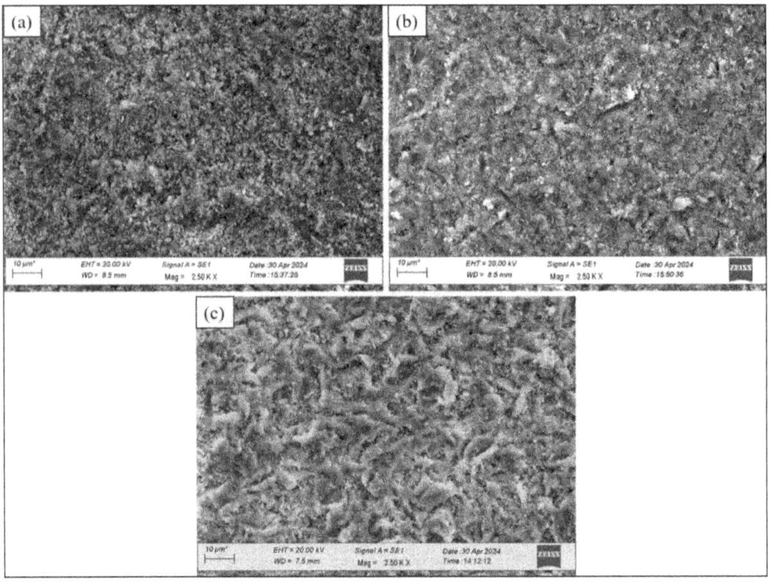

Fig. 5.15 Eroded surfaces of (a) SR-5, (b) SR-10 and (c) SR-15 AMCs at 72 m/s erodent velocity and 60° impingement angle

Fig. 5.16 illustrates the eroded surface profiles of the SR-5 composite at 100 m/s erodent velocity and impingement angles of 30°, 60° and 90°. The surface morphology clearly reveals that the pattern of crater creation by Al_2O_3 erodent changes with different impingement angles. Fig. 5.16a-c reveals that the size and depth of the craters developed at a 90° impingement angle are higher than those at 30° and 60°. This variation in surface morphology results from the differences in plastic deformation of the eroded samples. The eroded surfaces exhibit plastic deformation characterized by erosion channels and particle tearing, as shown in Fig. 16a and b respectively. In Fig. 16c, it is noted that at a 90° impingement angle, a crater pit forms because of the normal impact of erodent particles at high velocity.

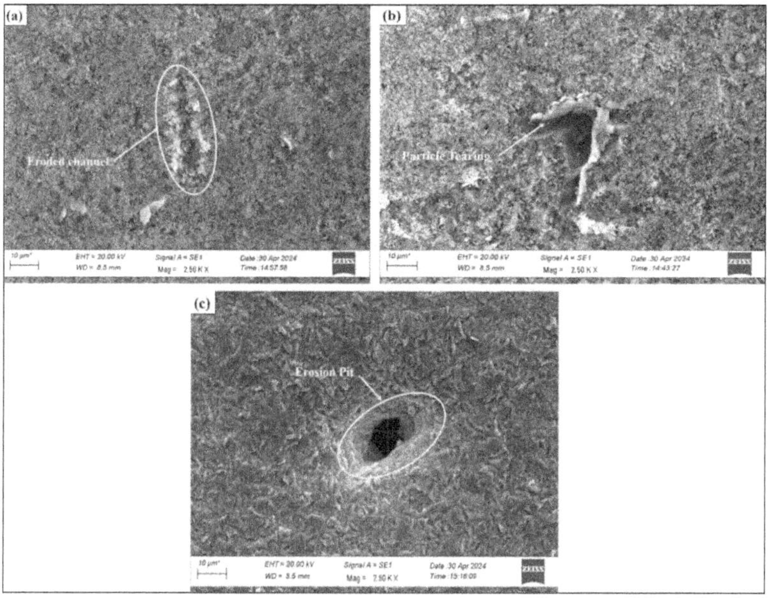

Fig. 5.16 Eroded surfaces of SR-5 AMC at (a) 30 °, (b) 60 ° and (c) 90 ° impingement angle and 100 m/s erodent velocity

CHAPTER 6

CONCLUSIONS AND SCOPE FOR FUTURE WORK

6.1 CONCLUSIONS

This investigation highlights the development of an affordable, light-weight substitute material for tribological applications like bearings, brake drums and sliding components of automobile. The impact of red mud and sillimanite particles on the physical, microstructural, mechanical and tribological properties of LM25 alloy matrix composite have been evaluated. All the composites successfully synthesized using stir casting method by varying wt. % of red mud and sillimanite particles. The main conclusions of this work can be summarized as follows:

6.1.1 Physical, Microstructural and Mechanical Properties

- The density of the composite increases as the amount of reinforcement is increased, due to the presence of high-density reinforcement particulates. The experimental density of the SR-15 composite is increased by 5.68% compared to the base alloy.

- The porosity of SR-5 composite (5.29%) is found to be maximum among all composites. The SR-10 and SR-15 composites have lower porosity than the SR-5 composites. Porosity was created due to gas entrapment during mixing, shrinkage and air bubbles entering the slurry during solidification.

- The thermal conductivity of composites shows a minor decrease as the reinforcement quantity increases. Thermal conductivity of the SR-15 composite is 3.13% lower than that of the LM25 alloy, due to the existence of voids and cavities. However, the thermal conductivity of synthesize AMCs exhibited remarkable heat dissipation capabilities in contrast to cast iron.

- The optical and SEM micrographs reveal a uniform distribution of the red mud and sillimanite particles throughout the LM25 alloy matrix. EDS analysis validates the existence of both reinforcements in the base alloy.

- The hardness of the composite increases with the wt.% of reinforcement. The SR-15 composite exhibits a 29.27% improvement in hardness when compared to LM25 alloy due to harder reinforcement particles.

- Ultimate tensile strength (UTS), yield strength and Youngs modulus increases with an increase in the weight percentage of both reinforcements. The SR-15 composite shows a 33.24% increase in UTS, 59.50% enhancement in yield strength and 10.66 % increase in the Youngs modulus compared to the base alloy. The enhancement in strength is due to the better bonding of reinforcement particles and Orowan strengthening. Conversely, as reinforcement particle amounts are increased, the ductility of AMCs decreases due to the restriction in dislocation movement.

- Fractography analysis shows that the LM25 alloy fractures in a ductile mode. The failure mode gradually shifts from ductile to brittle fracture with an increasing percentage of reinforcement. However, composites exhibit a mixed mode of failure (ductile and brittle), as evidenced by the decrease in the size of dimples in the micrographs.

6.1.2 Tribological Properties

- The quadratic regression model for wear rate is developed and analyzed using ANOVA based on BBD. The regression model indicates a strong relationship between input variables and wear rate. Applied normal load has the most significant impact on the wear rate, followed by sliding speed and reinforcement wt.%.

- The higher applied loads result in increased wear rates due to disturbed tribolayers and larger contact areas causing surface delamination and instability of MML under high stress.

- Sliding speed shows a parabolic effect on wear rate, with an initial increase due to frictional heat and a subsequent decrease after 1 m/s due to the formation of protective tribolayers or MML.

- The wear rate decreases with a higher concentration of reinforcement particles up to 10 wt.% reinforcement in base alloy, however, beyond it the wear rate increases. Among all composites, SR-10 shows the minimum wear rate because harder materials are more resistant to plastic deformation and abrasion, leading to a lower wear rate. However, beyond 10 wt.% reinforcement, the wear rate increases due to particle separation and ploughing effects, although a SR-15 composite has a lower wear rate than a SR-5 composite.

- The quadratic regression model for COF is developed and analyzed using ANOVA based on BBD. The regression model indicates a strong relationship between input variables and COF. Applied normal load has the most significant impact on the COF, followed by reinforcement weight percentage and sliding speed.

- The COF increases with applied load, which aligns with the fundamental law of friction stating that frictional force is proportional to the normal load.

- The COF reduces as the content of red mud and sillimanite particles increases within the base alloy because hard ceramic dispersoid acts as a protrusion on the contact surface, bearing maximum pressure and preventing abrasive particles from penetrating the pin.

- The COF increases with sliding speed up to 0.8 m/s, then decreases as sliding speed continues to rise, due to material softening at higher temperatures.

- Wear process parameters are successfully optimized using desirability analysis for minimum wear rate and COF. Reinforcement of 13.48 wt.%, sliding speed of 1.5 m/s and load of 10 N is found optimum for the wear rate of 0.01366 mm^3/m and COF 0.3615. Further the confirmation test results, conducted on these optimized parameters indicate the adequacy of the developed model.

- The wear track analysis of SR-15 composite under varying loads and fixed sliding speed reveals distinct wear mechanism. The composite exhibit abrasive wear mechanism at lower load while combination of adhesive and delamination wear mechanisms at higher load. Furthermore, the narrowing of grooves in SR-15 compared to SR-5 and SR-10 composites demonstrates improved abrasion resistance and enhanced load-carrying capabilities with higher reinforcement particle wt.%.

- The quadratic regression model for erosion rate is developed and analyzed using ANOVA based on BBD. The regression model indicates a strong relationship between input variables and erosion rate. Erodent velocity is the most significant factor among all the control parameters used for the erosion test, followed by the impingement angle and reinforcement wt.%.

- Increased erodent velocity results in a higher erosion rate due to greater particle impact energy and more frequent collisions.

- The erosion rate of the AMCs also increases with increasing impingement angle because at high impingement angles, material removal increases through cutting, indentation and crater formation.

- The SR-5 composite has minimum erosion rate due to the barrier created by reinforcement materials that resist the penetration and cutting action. The SR-10 and SR-15 AMCs experience increased erosion rates due to their brittle nature.

- Wear process parameters are successfully optimized using desirability analysis for minimum erosion rate. Reinforcement of 5 wt.%, erodent velocity of 1.5 m/s and impingement angle of 90° were found optimal for achieving an erosion rate of 0.0013883 gm/min. The confirmation test on optimized parameters signifies that the prediction tool has good accuracy.

- The SEM morphology of the eroded surfaces reveals plastic deformation, erosion pits, eroded channels and particle tearing which is corelated to the results obtained. The primary mechanisms of material removal were micro ploughing and micro cutting.

6.2 SCOPE FOR FUTURE WORK

In recent decades, the use of aluminium matrix composites has surged across various fields, including aerospace, defense, automotive, recreational equipment, office products, biomedical devices and engineering. Here are some suggestions that could be valuable for future research:

- The impact of diverse heat treatment process variables, including temperature and time, on the physical, microstructural, mechanical and tribological properties of AMCs can be studied.

- The AMCs can be investigated for various machining parameters and conditions like cutting speed, depth of cut, feed rate and cooling/lubrication conditions.

- The AMCs can be investigated for high-temperature performance and thermal stability under engine-specific conditions such as prolonged thermal exposure, cyclic loading at elevated temperatures and varying lubrication environments.

www.ingramcontent.com/pod-product-compliance
Lightning Source LLC
LaVergne TN
LVHW010737061126
851857LV00014B/1520

* 9 7 8 1 9 6 2 1 1 6 6 7 1 *